Studies in Smart Technologies

Advanced communication and computing technologies to develop intelligent and adaptive systems are usually referred to as smart technologies. Many of the sustainable development goals (e.g., SDG 4, SDG 6, SDG 7, SDG 8, SDG 9, SDG 11) cannot be achieved without developing smart technologies. The aim of the proposed book series, "Studies in Smart Technologies," is to provide a detailed and latest overview of the latest research findings in the field of smart technologies and allied areas. The series covers studies related to smart cities, IoT, smart sensors, AI, soft computing, and other related areas. The content published in the series will be an important reference to the policymakers, industries, researchers, practitioners, and students in computer science, electrical engineering, information technology, and aerospace engineering.

The series will publish monographs, edited volumes, textbooks and proceedings of important conferences, symposia, and meetings in the field of smart technologies and their applications.

Aritra Acharyya · Prasenjit Dey · Sujit Biswas
Editors

Real-World Applications and Implementations of IoT

Springer

Editors
Aritra Acharyya
Department of Electronics
and Communication Engineering
Kalyani Government Engineering College
Kalyani, West Bengal, India

Prasenjit Dey
Department of Computer Science
and Engineering
National Institute of Technology, Rourkela
Rourkela, Odisha, India

Sujit Biswas
University of London, Northampton Square
Cybersecurity and FinTech
London, UK

ISSN 3004-9881 ISSN 3004-989X (electronic)
Studies in Smart Technologies
ISBN 978-981-97-8626-8 ISBN 978-981-97-8627-5 (eBook)
https://doi.org/10.1007/978-981-97-8627-5

This Springer imprint is published by the registered company Springer Nature Singapore Pte Ltd.
The registered company address is: 152 Beach Road, #21-01/04 Gateway East, Singapore 189721, Singapore

Preface

The advent of the Internet of Things (IoT) has revolutionized how we interact with technology, bringing profound changes to various sectors. This book, *Real-World Applications and Implementations of IoT*, aims to explore these transformative impacts by delving into practical implementations and the benefits derived from IoT technologies. Our motivation in curating this collection stems from the recognition of IoT's potential to enhance efficiency, connectivity, and intelligence across diverse domains. With contributions from leading experts, each chapter provides insights into specific applications of IoT, illustrating how theoretical concepts translate into real-world solutions.

The book begins with an introductory chapter, setting the stage for subsequent discussions by providing a broad overview of IoT applications and implementations. From there, readers are guided through a variety of case studies and research findings. These include the integration of IoT in smart buildings, showcasing energy savings and performance optimization, and the role of IoT in resource management within cloud-based systems, highlighting the challenges and solutions for efficient resource allocation. One of the standout discussions is on smart street lighting systems, which not only improve energy conservation but also incorporate advanced technologies for urban infrastructure enhancement. Another notable contribution is the exploration of IoT in biomedical research, where advanced models like BERT are utilized for sequential text classification, exemplifying the intersection of IoT and artificial intelligence. Further chapters address innovative IoT applications such as Wi-Fi enabled communications with a focus on energy efficiency, drone-based systems for agricultural advancements, and IoT implementations in predictive maintenance, health monitoring, and beyond.

Our goal with this book is to provide a comprehensive resource that encapsulates the vast potential and versatility of IoT. We hope it serves as a valuable guide for researchers, practitioners, and anyone interested in understanding the impactful ways IoT is shaping our world. We extend our gratitude to all contributing authors for their invaluable insights and to our readers for embarking on this journey through the evolving landscape of IoT.

Kalyani, West Bengal, India — Aritra Acharyya
Rourkela, Odisha, India — Prasenjit Dey
London, UK — Sujit Biswas

About This Book

This book offers an in-depth exploration of practical applications and implementations of the Internet of Things (IoT), presenting a wide spectrum of insights from foundational concepts to advanced technologies. Key highlights include the integration of IoT in smart buildings for optimized performance and energy savings, and the challenges and solutions in resource management within IoT-cloud environments. The book delves into IoT-enabled smart street lighting systems, biomedical applications using advanced models like BERT for sequential text classification, and energy-efficient Wi-Fi enabled IoT communications. It also covers drone-based agricultural innovations and various real-world applications such as predictive maintenance and health monitoring. This comprehensive resource bridges the gap between theoretical concepts and practical implementations of IoT, serving as an invaluable guide for researchers, practitioners, and students interested in understanding the transformative potential of IoT in our increasingly connected world.

Contents

Editors and Contributors

About the Editors

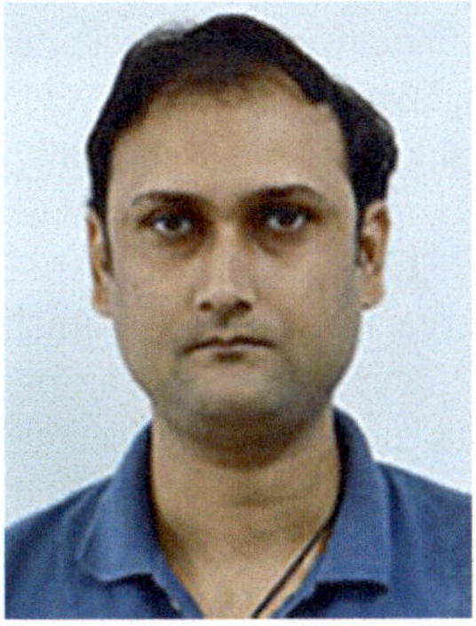

Dr. Aritra Acharyya is currently working at the Department of Electronics and Communication Engineering, Kalyani Government Engineering College, Department of Electronics and Communication Engineering, Kalyani, West Bengal - 741235, India, as an Assistant Professor. He was born in 1986. He received B.E. and M.Tech. degrees from IIEST, Shibpur, India, and the Institute of Radio Physics and Electronics, University of Calcutta, India, in the years 2007 and 2010 respectively. Finally, he obtained Ph.D. degree from the Institute of Radio Physics and Electronics, University of Calcutta, in the year 2016. His research interests are high-frequency semiconductor devices, nano-structures, semiconductor physics, transport phenomena, quantum mechanics, optoelectronics, IoT, machine learning, etc. He has published 95 research papers in peer-reviewed national and international journals, 64 research papers in national and international conference proceedings, and several book chapters. He also authored and edited 07 and 06 number of books respectively.

Dr. Prasenjit Dey received the B.Tech. Degree in computer science and engineering from the West Bengal University of Technology, Kolkata, India, in 2010, and the M.Tech. Degree in computer science and engineering from the National Institute of Technology Durgapur, Durgapur, India, in 2012. He was an Associate Innovator with Nivio Technologies, Gurgaon, India, in 2012. He was a hardware graphics designer at Intel from 2017 to 2018. He received his Ph.D. degree in computer science and engineering from the National Institute of Technology Durgapur, Durgapur, India, in 2018. He was an Assistant Professor in the computer science and engineering at Cooch Behar Government Engineering College, Cooch Behar, India. He joined the National Institute of Technology Rourkela, Odisha, India, as an Assistant Professor in the Department of Computer Science in 2023 and is currently working. His current research interests include artificial neural networks, pattern recognition, machine learning, etc. He has many research publications in internationally reputed journals like IEEE Transactions on Systems, Man, and Cybernetics, IEEE Transactions on Artificial Intelligence, IEEE Internet of Things Journal Impact, Biomedical Signal Processing and Control, Pattern Recognition, etc. He has more than 15 research publications in internationally reputed edited books and international conference proceedings.

Dr. Sujit Biswas is a distinguished blockchain researcher and academic with a profound impact on the field of distributed ledger technology. He is a Lecturer (equivalent to Assistant Professor) in Cybersecurity and FinTech, Computer Science Department, City, University of London, UK, and Research Associate (Honorary) at CBT, University College London (UCL). Prior to this position, he was a Lecturer in Computer Science and Digital Technologies at the University of East London; a research fellow in blockchain and AI at the University of Surrey's Centre for Vision, Speech, and Signal Processing (CVSSP). He also served as an Assistant Professor in the Department of Computer Science and Engineering at Faridpur Engineering College, Dhaka University, Bangladesh. He holds a Ph.D. in Computer Science and Technology from the Beijing Institute of Technology, China, and has dedicated his career to exploring the theoretical foundations and practical applications of blockchain technology. Dr. Sujit's research focuses on consensus algorithms, scalability solutions, and privacy-enhancing techniques within blockchain systems. His innovative work has been published in numerous prestigious journals and presented at international conferences, earning her recognition as a thought leader in the blockchain space. For instance, the IEEE Internet of Things Journal, IEEE Transactions on Big Data, IEEE Transactions on Engineering Management, IEEE Transactions on Service Management, Computer Magazine, ACM Computing Surveys, etc. In addition to his research contributions, Dr. Sujit Biswas is a passionate educator and mentor, inspiring the next generation of Blockchain enthusiasts, Federated Machine Learning, Deep Learning, Machine Learning domains, Internet of Things, etc. Furthermore, Dr. Sujit is actively involved in collaborative projects with both industry and government entities, working to bridge the gap between academia and real-world blockchain applications.

Contributors

Pratik Acharjee University of Engineering and Management, New Town, Kolkata, India

Aritra Acharyya Department of Electronics and Communication Engineering, Kalyani Government Engineering College, Kalyani, Nadia, West Bengal, India

Harishitha Chowdary Alapati SRM University, Amaravati, Andhra Pradesh, India

Erin Tabassum Arpa Department of Electrical and Electronic Engineering, Faridpur Engineering College, Faridpur, Bangladesh

Abhijit Biswas Department of Electronics and Communications Engineering, Supreme Knowledge Foundation Group of Institution, Hooghly, West Bengal, India

Sujit Biswas Cybersecurity and FinTech, Department of Computer Science, School of Science and Technology, City University of London, London, UK

Aritrya Chatterjee University of Engineering and Management, New Town, Kolkata, India

Paromita Das Department of Electronics and Communication Engineering, Siliguri Institute of Technology, Siliguri, West Bengal, India

Debashis De Department of Computer Science and Engineering, Centre of Mobile Cloud Computing, Maulana Abul Kalam Azad University of Technology, Nadia, West Bengal, India

Prasenjit Dey Department of Computer Science and Engineering, National Institute of Technology, Rourkela, Odisha, India

Arpan Deyasi Department of Electronics and Communication Engineering, RCC Institute of Information Technology, Kolkata, India

Sayani Dhali Department of Electronics and Communications Engineering, Supreme Knowledge Foundation Group of Institution, Hooghly, West Bengal, India

Soumyadip Dhar Department of Information Technology, RCC Institute of Information Technology, Kolkata, India

Sai Harshitha Dhulipalla SRM University, Amaravati, Andhra Pradesh, India

Aditya Dudugu SRM University, Amaravati, Andhra Pradesh, India

Swetha Ghanta SRM University, Amaravati, Andhra Pradesh, India

Dia Ghosh Department of Electronics and Communication Engineering, Siliguri Institute of Technology, Siliguri, West Bengal, India

Mridul Ghosh Department of Computer Science, Shyampur Siddheswari Mahavidyalaya, Ajodhya, Shyampur, West Bengal, India

Jayabrata Goswami Netaji Subhas Open University, Saltlake, Kolkata, India

Md. Ariful Islam Department of Electrical and Electronic Engineering, Faridpur Engineering College, Faridpur, Bangladesh

Akash Jana Department of Computer Science, Surendranath College, Kolkata, West Bengal, India

Arun Kumar Maiti Department of Mathematics, Shyampur Siddheswari Mahavidyalaya, Ajodhya, Shyampur, West Bengal, India

Kanishka Majumder Academy of Technology (AOT), Hooghly, West Bengal, India

Purbayan Majumder Accenture Security, Gurugram, Haryana, India

Biplab K. Mandal Department of Information Technology, Asansol Engineering College, Asansol, West Burdwan, West Bengal, India

Partha Mandal Department of Electrical and Electronic Engineering, Faridpur Engineering College, Faridpur, Bangladesh

Avirup Manjumder University of Engineering and Management, New Town, Kolkata, India

Sravya Meda SRM University, Amaravati, Andhra Pradesh, India

Sharmin Sultana Mim Department of Electrical and Electronic Engineering, Faridpur Engineering College, Faridpur, Bangladesh

Anwesha Mukherjee Department of Computer Science, Mahishadal Raj College, Purba Medinipur, West Bengal, India

Poulomi Mukherjee Department of Computer Science and Engineering, National Institute of Technology, Durgapur, India

Shaik Tahseen Nishat SRM University, Amaravati, Andhra Pradesh, India

Ashok Kumar Pradhan SRM University, Amaravati, Andhra Pradesh, India

Subham Pramanik Academy of Technology (AOT), Hooghly, West Bengal, India

Anindita Raychaudhuri Department of Computer Science, Sarojini Naidu College for Women, Kolkata, West Bengal, India

Annapurna Roy Department of Computer Science, Shyampur Siddheswari Mahavidyalaya, Ajodhya, Shyampur, West Bengal, India

Hiranmoy Roy Department of Information Technology, RCC Institute of Information Technology, Kolkata, India

Poly Saha Department of Plant Pathology, Bidhan Chandra Krishi Viswavidyalaya, Bardhaman, India

Rajdeep Saha Department of Electronics and Communications Engineering, Supreme Knowledge Foundation Group of Institution, Hooghly, West Bengal, India

Priya Sarkar Department of Computer Science, Surendranath College, Kolkata, West Bengal, India

Saradwat Sen Department of Electronics and Communications Engineering, Supreme Knowledge Foundation Group of Institution, Hooghly, West Bengal, India

Md Tarikul Islam Sheam Department of Electrical and Electronic Engineering, Faridpur Engineering College, Faridpur, Bangladesh

Babul P. Tewari Department of Computer Science and Engineering, Ghani Khan Choudhury Institute of Engineering and Technology, Malda, West Bengal, India

Preetam Vallabhaneni SRM University, Amaravati, Andhra Pradesh, India

Chapter 1
Introduction to Real-World Applications and Implementations of IoT

Aritra Acharyya, Prasenjit Dey, and Sujit Biswas

Abstract The Internet of Things (IoT) is a transformative technology that has become integral to various sectors, including energy conservation, security, agriculture, mining, healthcare, and environmental protection. This book, "Exploring the Cutting-Edge of the Internet of Things," delves into the latest advancements and applications of IoT, highlighting its potential to optimize energy consumption, enhance security, improve agricultural practices, ensure mining safety, and protect the environment. By combining IoT with artificial intelligence (AI), the book explores new possibilities and advancements that push the boundaries of innovation. Originating from the Massachusetts Institute of Technology in 1999, IoT has grown into a vast network interconnecting physical and virtual entities through standardized communication protocols, enabling intelligent interactions such as positioning, tracking, identification, perception, monitoring, and management across diverse networks. This book caters to a diverse audience, from beginners with a foundational understanding of science to seasoned experts in engineering and technology, offering valuable insights and knowledge for researchers, practitioners, and students aiming to stay at the forefront of IoT and AI innovations.

Keywords Internet of Things (IoT) · Artificial intelligence (AI) · Energy conservation · Smart agriculture · Environmental protection · Industrial automation

A. Acharyya (✉)
Department of Electronics and Communication Engineering, Kalyani Government Engineering College, Kalyani, Nadia, West Bengal 741235, India
e-mail: ari_besu@yahoo.co.in

P. Dey
Department of Computer Science and Engineering, National Institute of Technology, Rourkela Sector 1, Rourkela, Odisha 769008, India
e-mail: deyp@nitrkl.ac.in

S. Biswas
Cybersecurity and FinTech, Department of Computer Science, School of Science and Technology, City University of London, Northampton Square, London EC1V 0HB, UK
e-mail: sujit.biswas@city.ac.uk

A. Acharyya et al. (eds.), *Real-World Applications and Implementations of IoT*, Studies in Smart Technologies, https://doi.org/10.1007/978-981-97-8627-5_1

1 Introduction

In today's ever-changing world, it has become imperative to address critical issues such as energy conservation, security, agricultural advancements, mining safety, healthcare, and environmental protection. This book, "Real-World Applications and Implementations of IoT" embarks on a journey to explore the state-of-the-art solutions provided by the Internet-of-Things (IoT). These solutions span a wide range, from optimizing energy consumption to safeguarding our environment, including the conservation of forests and wildlife. The book takes a deep dive into this transformative technology, offering a comprehensive analysis, detailed descriptions, and in-depth discussions of recently developed IoT applications. With a global perspective and a strong focus on cutting-edge research, the book uniquely combines IoT with artificial intelligence (AI), illuminating emerging possibilities and advancements. The Internet of Things, which traces its roots back to its inception at the Massachusetts Institute of Technology in 1999, has evolved into a vast network. It interconnects physical and virtual entities through standardized communication protocols, embracing a plethora of sensors and actuators, each possessing unique attributes [1–10]. These components enable intelligent interactions encompassing positioning, tracking, identification, perception, monitoring, and management across diverse networks, transcending temporal and spatial boundaries. IoT has seamlessly integrated itself into various facets of our lives, from the realms of healthcare and smart-homes to industrial control and agriculture [11–20].

Designed to cater to a diverse audience, from those with a foundational understanding of science to seasoned engineering and technology experts, this book serves as an invaluable resource for engineering students and science master's programs. Additionally, it provides a rich source of knowledge for researchers striving to stay at the forefront of IoT and AI innovations. Readers are invited to embark on a journey through the captivating landscape of IoT and discover its remarkable potential to shape a brighter and more sustainable future. This book focuses on showcasing some of the latest developments in various fields related to the IoT. It will feature comprehensive discussions of recently developed and upcoming IoT solutions, spanning areas such as energy conservation, social security, agricultural progress, mining security, and environmental conservation, including forest and wildlife preservation. Such an extensive coverage of topics is currently lacking in the available literature in the form of a book. This proposed work seeks to delve deeply into these areas, emphasizing theoretical, methodological, well-established, and validated empirical research, alongside detailed experimental and application aspects. The book is set to encompass 15 chapters, complete with over 100 color illustrations and pertinent photographs to aid in the comprehension of the subject matter. The editors spearheading this project are experts in the field, boasting a proven track record in their respective areas of expertise.

One notable feature of this proposed book is its unique positioning in the market. While there are fundamental books that delve into related fields, they often require readers to refer to a multitude of different books and sources for a comprehensive understanding of the subject matter. The book stands out by providing an all-encompassing view of the topic, backed by state-of-the-art technological reviews. Existing books either lack experimental depth or application aspects or focus solely on one specific topic rather than offering an overview of the broader spectrum. This book is designed to fill this void by offering a comprehensive understanding of the entire field, making it more accessible to professionals, students, and researchers in the fields of electronics, communication, computer science, electrical engineering, and related subjects. In summary, this book aims to serve as a valuable reference for those looking to explore the depths of IoT and its applications.

2 Overview of the Book

The Internet of Things (IoT) represents a paradigm shift in the way technology intersects with daily life, industry, and various sectors of the economy. This book, "Real-World Applications and Implementations of IoT," delves into this transformation by exploring a myriad of practical applications and the substantial benefits derived from IoT technologies. Each chapter has been meticulously crafted by leading experts to provide comprehensive insights into specific domains where IoT is making a significant impact.

This Chapter (Overview of IoT Applications and Implementations): The opening chapter sets the stage by providing a broad overview of IoT. It introduces the fundamental concepts, the evolution of IoT technologies, and their significance in the modern world. The chapter outlines the structure of the book and previews the various domains covered in subsequent chapters. It serves as a primer for readers, ensuring they have a foundational understanding of IoT before delving into specific applications.

Chapter 2 (IoT in Smart Buildings): This chapter explores the integration of IoT in smart buildings, emphasizing how IoT technologies contribute to energy savings and performance optimization. It discusses the role of sensors and analytics in real-time monitoring and control of building systems, including HVAC, lighting, and security. Case studies highlight successful implementations of smart building technologies, showcasing tangible benefits such as reduced energy consumption and enhanced occupant comfort.

Chapter 3 (Resource Management in IoT-Cloud Systems): Chapter 3 addresses the challenges and solutions associated with resource management in IoT-cloud environments. It examines the critical need for efficient resource allocation to handle the vast amounts of data generated by IoT devices. The chapter discusses latency reduction, real-time processing, and innovative algorithms that enhance the performance of IoT-cloud systems. Practical examples illustrate how these solutions are applied in real-world scenarios.

Chapter 4 (Smart Street Lighting Systems): This chapter delves into IoT-enabled smart street lighting systems, which offer significant advancements in urban infrastructure. It details how these systems use sensors and connectivity to provide automated illumination, dynamic dimming, remote control, and real-time monitoring. The chapter highlights the benefits of smart street lighting, such as energy conservation, reduced maintenance costs, and improved urban safety. Real-world implementations are presented to underscore the effectiveness of these technologies.

Chapter 5 (IoT in Biomedical Research): Chapter 5 explores the intersection of IoT and artificial intelligence in biomedical research. It focuses on advanced models like BERT (Bidirectional Encoder Representations from Transformers) for sequential text classification, demonstrating how IoT can enhance data retrieval and classification in biomedical databases. The chapter provides case studies showing the practical applications of these technologies in medical research, improving the accuracy and efficiency of data analysis.

Chapter 6 (Energy-Efficient IoT Communications): This chapter investigates energy-efficient IoT communications, particularly focusing on Wi-Fi enabled devices. It discusses the challenges of power consumption in battery-operated IoT devices and presents strategies to optimize energy usage. The chapter includes practical insights into the design and implementation of energy-efficient communication protocols, ensuring that IoT devices maintain functionality and longevity.

Chapter 7 (IoT in Agriculture): Chapter 7 highlights innovative IoT applications in agriculture, specifically the use of drone-based systems for detecting forest medicinal plants. It explores how IoT technologies can revolutionize agricultural practices by providing precise, real-time data for crop management and monitoring. The chapter presents case studies demonstrating the impact of IoT on agricultural efficiency and productivity, showcasing how technology can address traditional farming challenges.

Chapter 8 (Predictive Maintenance Using IoT): This chapter examines the role of IoT in predictive maintenance, a critical application in various industries. It explains how IoT sensors and analytics can predict equipment failures before they occur, reducing downtime and maintenance costs. The chapter provides examples from industries such as manufacturing and transportation, where predictive maintenance has led to significant improvements in operational efficiency.

Chapter 9 (Health Monitoring with IoT): Chapter 9 discusses the applications of IoT in health monitoring, emphasizing its potential to transform healthcare delivery. It explores how IoT devices can monitor patients' vital signs in real-time, enabling early detection of health issues and personalized treatment. The chapter includes case studies illustrating the benefits of IoT in improving patient outcomes and streamlining healthcare processes.

Chapter 10 (Intelligent Surveillance Systems): This chapter explores the use of IoT in intelligent surveillance systems, which enhance security and safety in various environments. It discusses how IoT technologies enable real-time monitoring, automated threat detection, and response. The chapter presents case studies of intelligent surveillance implementations in public spaces, commercial buildings, and critical infrastructure, highlighting the increased effectiveness and efficiency of these systems.

Chapter 11 (IoT in Industrial Automation): Chapter 11 delves into the applications of IoT in industrial automation, where it drives significant advancements in manufacturing processes. It discusses how IoT sensors and connectivity enable real-time monitoring and control of industrial equipment, leading to improved productivity and reduced operational costs. The chapter provides examples of IoT-driven automation in various industries, showcasing the transformative impact of these technologies.

Chapter 12 (Smart Grid and IoT): This chapter examines the integration of IoT in smart grid systems, which are essential for modernizing the electricity grid. It explores how IoT technologies facilitate real-time monitoring, efficient energy distribution, and improved grid reliability. The chapter includes case studies of smart grid implementations, demonstrating the benefits of IoT in enhancing energy management and sustainability.

Chapter 13 (IoT in Transportation and Logistics): Chapter 13 focuses on the applications of IoT in transportation and logistics, where it enhances efficiency and safety. It discusses how IoT technologies enable real-time tracking of vehicles and goods, optimize routes, and improve fleet management. The chapter provides examples of IoT implementations in logistics and transportation, highlighting the operational and economic benefits achieved through these innovations.

Chapter 14 (Environmental Monitoring Using IoT): This chapter explores the role of IoT in environmental monitoring, which is crucial for addressing global environmental challenges. It discusses how IoT sensors and networks can monitor air and water quality, track wildlife, and detect environmental changes. The chapter presents case studies demonstrating the impact of IoT on environmental conservation and management efforts.

Chapter 15 (Future Trends and Challenges in IoT): The final chapter looks ahead to the future of IoT, discussing emerging trends and potential challenges. It explores advancements in IoT technologies, such as edge computing, 5G connectivity, and artificial intelligence integration. The chapter also addresses the challenges of scalability, security, and privacy in IoT implementations, providing insights into how these issues can be mitigated. It concludes with a discussion on the potential future impact of IoT on various sectors and the ongoing evolution of this transformative technology.

This comprehensive book aims to bridge the gap between theoretical concepts and practical implementations of IoT, offering valuable insights for researchers, practitioners, and students. By showcasing a wide range of real-world applications, it highlights the transformative potential of IoT in various domains, underscoring its role in shaping a more connected and efficient world.

3 Conclusion

The Internet of Things (IoT) stands at the forefront of technological innovation, driving significant advancements across multiple domains. This book provides a comprehensive exploration of IoT's transformative potential, presenting state-of-the-art solutions and applications that span energy conservation, security, agriculture, mining, healthcare, and environmental protection. The integration of IoT with artificial intelligence (AI) further enhances its capabilities, enabling intelligent interactions and decision-making processes that transcend traditional boundaries. As IoT continues to evolve, its impact on various sectors becomes increasingly profound, offering solutions that optimize energy use, enhance security, improve agricultural efficiency, ensure mining safety, and protect the environment. This book serves as an invaluable resource for engineering students, science master's programs, researchers, and practitioners, providing detailed discussions, comprehensive analyses, and real-world case studies that illuminate the practical applications and theoretical foundations of IoT. By bridging the gap between theoretical concepts and practical implementations, the book underscores the role of IoT in shaping a more connected, efficient, and sustainable future. As we look to the future, emerging trends such as edge computing, 5G connectivity, and AI integration will further drive the evolution of IoT, presenting new opportunities and challenges that require innovative solutions and continuous research. This comprehensive guide offers a roadmap for understanding and leveraging the potential of IoT, highlighting its significance in the ongoing journey toward a smarter, more interconnected world.

References

1. Atzori L, Iera A, Morabito G (2010) The internet of things: a survey. Comput Netw 54(15):2787–2805
2. Vermesan O, Friess P, Guillemin P, Gusmeroli S, Sundmaeker H, Bassi A (2011) Internet of things strategic research roadmap. River Publishers
3. Gubbi J, Buyya R, Marusic S, Palaniswami M (2013) Internet of things (IoT): a vision, architectural elements, and future directions. Futur Gener Comput Syst 29(7):1645–1660
4. Borgia E (2014) The internet of things vision: key features, applications and open issues. Comput Commun 54:1–31
5. Shi W, Cao J, Zhang Q, Li Y, Xu L (2011) Edge computing: vision and challenges. IEEE Internet Things J 3(5):637–646
6. Hasan KAMM, Hossain MS, Muhammad G, Alelaiwi A (2018) Real-time big data processing in mobile cloud computing: current trends and future directions. IEEE Cloud Computing 5(2):44–51
7. Gia TN, Jiang M, Rahmani R, Westerlund T (2015) Internet of things for smart cities: a trust model from the security perspective. IEEE Internet Things J 3(6):1190–1198
8. Tao J, Yang C, Wang Q (2016) An IoT-oriented data storage framework in cloud computing platform. IEEE Internet Things J 3(3):349–359
9. Baccarelli E, Makrakis D, Pizlo F (2015) IoT-based environmental monitoring and actuation for the water domain. IEEE Internet Things J 2(2):97–106

10. Jin X, Zhang Z (2014) An IoT-oriented data storage framework in cloud computing platform. IEEE Internet Things J 2(6):482–489
11. Al-Fuqaha A, Guizani M, Mohammadi M, Aledhari M, Ayyash M (2015) Internet of things: a survey on enabling technologies, protocols, and applications. IEEE Commun Surv Tutor 17(4):2347–2376
12. Cavalcanti D, Loureiro AAF, Boukerche A (2016) Vehicular ad hoc networks: a new challenge for localization-based systems. IEEE Internet Things J 3(4):515–526
13. Yassein MB, Al-Zou'bi AS, Almazo M, Guizani M (2018) IoT-based smart cities: a survey. IEEE Internet Things J 5(6):3913–3932
14. Fan Y, Wang W, Sun J, Qian J, Maharjan S (2019) Edge caching and computing for internet of things: a review, use cases, and future trends. IEEE Internet Things J 6(2):2171–2188
15. Ray PP (2016) A survey of IoT cloud platforms. Futur Gener Comput Syst 56:684–700
16. Ashton K (2009) That internet of things thing. In the real world, things matter more than ideas. RFID J 22(7)
17. Sheng Z, Yang S, Yu Y, Vasilakos AV, McCann JA, Leung KK (2013) A survey of the research on future networks and its integration with cloud computing. IEEE Commun Surv Tutor 15(3):1017–1044
18. Vermesan O, Friess P (eds) (2013) Internet of things: global technological and societal trends. River Publishers
19. Vermesan O, Friess P (eds) (2014) Building the hyperconnected society: internet of things research and innovation value chains, ecosystems, and markets. River Publishers
20. Yassein MB, Otoom M (2016) The internet of things: applications and challenges in technology and standardization. CRC Press

Chapter 2
Implementation of Smart Building Using Internet of Things (IoT)

Kanishka Majumder, Subham Pramanik, and Jayabrata Goswami

Abstract Building owners and operators each over the world are looking for new ways to streamline the operations to get smart structure systems that are largely optimized and useful. Owners and drivers may need to invest between hundreds of thousands and millions of dollars to add smart structure technologies, but such systems certainly help to improve energy savings, reduce energy conservation costs, and improve overall structure performance. Investments in smart structure technology typically pay themselves within one to two times by providing energy savings and a conservation advantage. Utilizing the right Internet-of-Things (IoT), detectors, demesne gateway, distributed control, and pallet analytics can be used. Smart structures have the capability to collect, dissect, and sort the structure data snappily to offer real-time energy and performance optimization. A smart structure does not stop at the four walls that compass it. Utmost of the smart structure results can interact with the external surroundings also. Externalities include variables similar as rainfall prognostications, residency, business events, transportation, etc., metropolizes thick and consume further energy. With the adding population, new civic growth is the driving demand for the structures and energy use. This creates numerous further openings for the structures to play a bigger part in a smarter earth—a system that tells us how to conserve energy. Smart structures are an essential element of smart metropolises, offering innovative results to enhance energy effectiveness, comfort, and security. The IoT plays a pivotal part in transubstantiating traditional structures into intelligent architectures. This exploration composition explores the perpetration of smart structures using IoT technologies, fastening on the integration of detectors, selectors, and communication networks. We bandy the benefits, challenges, and unborn prospects

K. Majumder (✉) · S. Pramanik
Academy of Technology (AOT), G.T. Road (Adisaptagram), Aedconagar, Hooghly, West Bengal 712121, India
e-mail: kanishka.majumder@aot.edu.in

S. Pramanik
e-mail: subham.pramanik@aot.edu.in

J. Goswami (✉)
Netaji Subhas Open University, DD-26, Sector 1, Salt Lake, Kolkata 700064, India
e-mail: goswamijayabrata@gmail.com

A. Acharyya et al. (eds.), *Real-World Applications and Implementations of IoT*, Studies in Smart Technologies, https://doi.org/10.1007/978-981-97-8627-5_2

of IoT in smart structures, along with case studies and recommendations for effective perpetration.

Keywords IoT · Smart buiding · Smart garden · Smart city · Smart structure

1 Introduction

IoT is a rapidly expanding technology that is changing the way we live, work, and play. IoT is the network of physical objects, vehicles, buildings, and other objects that use sensors, software, and other technologies to communicate and exchange information with each other over the internet. One of the most interesting and promising uses of IoT is in smart buildings [1–5]. Smart buildings are buildings that utilize IoT innovation to computerize and optimize their operations, move forward vitality productivity, upgrade tenant consolation and security, and make unused administrations and commerce openings. By interfacing different building frameworks and gadgets, such as HVAC, lighting, security, and vitality administration, smart buildings can make a more coordinate and responsive environment that can adjust to the requirements and inclinations of tenants. The implementation of smart buildings using IoT technology requires a holistic and interdisciplinary approach that involves various stakeholders, such as architects, engineers, building owners, facilities managers, and technology providers. The process of implementing smart buildings involves several stages, from planning and design to installation, commissioning, and operation.

The first stage is the planning and design phase, where the goals, requirements, and constraints of the smart building are defined. This involves identifying the stakeholders, defining the functional and technical requirements, selecting the appropriate IoT technologies and platforms, and creating a detailed design and implementation plan. The second stage is the installation and commissioning phase, where the IoT devices and systems are installed, tested, and integrated. This involves selecting and procuring the IoT hardware and software, installing the sensors, actuators, and gateways, configuring the network and security settings, and testing the system functionality and performance. The third stage is the operation phase, where the smart building is managed, monitored, and optimized. This involves setting up the user interfaces and dashboards, defining the control logic and algorithms, monitoring the system performance and energy consumption, and providing ongoing support and maintenance.

The implementation of smart buildings using IoT technology offers several benefits, such as increased energy efficiency, improved occupant comfort and productivity, reduced maintenance costs, enhanced security and safety, and new revenue streams. However, it also poses several challenges, such as data privacy and security, interoperability and standardization, cost and complexity, and organizational and regulatory barriers. To overcome these challenges and ensure the successful implementation of smart buildings using IoT technology, it is essential to adopt a collaborative and iterative approach that involves all the stakeholders, leverages best practices and standards, and focuses on user-centered design and value creation. By doing so, smart buildings can become a key driver of sustainable and resilient urban development, contributing to the well-being and prosperity of cities and communities.

2 Why IoT

Early experimentation and course of action of IoT frameworks begun with meddling mechanical devices. These days, the IoT vision has amplified to relate everything from mechanical devices to normal objects. The sorts of items amplify from gas turbines to cars and utility meters. It can as well consolidate living creatures such as plants, private animals, and individuals. For outline, the Bovine Taking after Wander in Essex businesses information collected from radio-location names to screen dairy animals' contamination and animals' conduct. Wearable computers and computerized prosperity gadgets similar to the Nike+ Fuel band and Fitbit are outlines of how people meddle inside the IoT environment. Cisco has extended the definition of IoT to incorporate the complete Web of Things (IoE), which joins people, places, objects, and things. In a general sense, anything you'll connect a sensor to and interface to can take portion inside the advanced related situations.

Connectable physical objects have one or more sensors. Each sensor screens certain conditions such as area, vibration, movement, and temperature. In IoT, these sensors interface to each other and to frameworks that can get it or speak to data from the sensor inputs. These sensors give unused data to venture frameworks and individuals. Within the past, individuals connecting with individuals and machines. Envision in the event that all devices may communicate. IoT-enabled objects share data around their state and their environment with individuals, computer program frameworks, and other machines. This data can be shared in genuine time or collected and shared at interims. Within the future, everything will have a digital character and network, which implies you'll be able to distinguish, track, and connect with objects. With progresses in all zones of life, it is vital to make modern standards to back development and increment client esteem. In a fast-paced trade environment. Shrewd Building Arrangements are an intelligent automated framework that makes a comfortable environment that makes a difference in strides building performance and energy proficiency.

The effect of savvy buildings on our lifestyle is awesome. Envision a domestic or office that recognizes you as a person, whether you are a property holder, occupant, worker, or provider. Envision an office that recognizes who enters the building, what physical get to they require, and what data they require. The house or office knows your inclinations for light, temperature, and room sort. Shrewd buildings make life less demanding and more helpful. The integration of sensors and remote arrangements is necessary to decrease working costs and spare vitality. Keen buildings too offer assistance to screen the utilization and lifecycle of your gadgets. In brief, keen building arrangements make spaces more intelligent, more secure, and more beneficial. IoT information is distinctive from conventional information handling. Information may be little in measure and sent regularly. The number of devices associated to organize, i.e. hubs, is additionally larger in the Web of Things than in conventional individual computers. Machine-to-machine communication and gadget arrange insights permit businesses to computerize certain center operations without depending on centralized or cloud-based applications and administrations. These attributes give openings to gather an assortment of information, but to show challenges in the plan of a fitting information organization and information security.

3 Evolution of Smart Buildings with IoT

Smart buildings have a rich history, dating back to the late 1800s with the invention of the thermostat. However, the modern era of smart buildings is characterized by the integration of IoT devices, cloud computing, and analytics driven by machine learning. These technologies have enabled buildings to automatically control operations based on real-time variables, optimizing energy use, space utilization, and environmental impact.

Benefits of Smart Buildings Enabled by IoT:

(a) **Insights**: IoT technologies enable continuous data collection and analysis, providing critical insights into building conditions and performance.
(b) **Environmental impact**: Efficiency-focused smart technology minimizes carbon emissions, contributing to environmental sustainability.
(c) **Comfort**: Intelligent automation of systems like lighting, heating, and air quality enhances occupant comfort.
(d) **Health and safety**: IoT devices monitor indoor air quality, control access, and maintain healthy conditions, ensuring occupant safety.
(e) **Maintenance**: Predictive maintenance through analytics software helps identify and address issues before they escalate.
(f) **Reduced energy costs**: Smart building systems optimize energy use, reducing costs and promoting sustainability.
(g) **Enhancing energy performance**: Detailed monitoring of energy consumption allows for targeted energy-saving measures.
(h) **Compliance with regulations**: IoT technologies facilitate compliance with energy regulations and standards, improving building energy performance.
(i) **Optimizing maintenance**: IoT sensors enable predictive maintenance, preventing structural defects and ensuring building safety.
(j) **Space utilization**: Smart occupancy monitoring optimizes space usage, especially crucial in the context of changing work patterns post-pandemic.
(k) **Automation and optimization**: Data collection and analysis drive efficiencies, cost savings, and informed decision-making in building management.

(l) **Modernizing without heavy construction**: IoT enables cost-effective modernization of existing building systems by adding sensors without extensive renovations.
(m) **Recovering data wirelessly**: Wireless IoT sensors are advantageous in situations where wired connections are impractical, providing centralized data access.
(n) **Supervising, anticipating, and adapting**: IoT applications in real estate optimize building performance, anticipate maintenance needs, and adapt to changing requirements.

4 Various Implementation of IoT for Smart Building

Intelligent parking can be defined as the use of advanced technologies for efficient operation, monitoring, and management of parking within the urban traffic strategy. The global market for smart parking systems reached $93.5 million, with the United States accounting for 46% of the market, offering strong growth opportunities for companies offering services both in the United States and abroad. Several technologies form the basis of smart parking solutions, including vehicle sensors, wireless communications, and data analytics. Smart parking is also made possible by innovations such as customer service smartphone applications, mobile payments, and car navigation systems. At the heart of the smart parking concept is the ability to receive, collect, analyze, distribute, and act on parking information. Increasingly, this information is obtained from smart devices in real time, which allows both parking attendants and drivers to optimize the use of parking spaces [6].

The common purpose of all car parks is to provide facilities for safety, accessibility, and maintenance. Therefore, the following points must be kept in mind (1) the location of the parking lot must be easily accessible. (2) The entrance to the parking lot must be easy to find. (3) There must be plenty of stopping spaces and there must be sufficient space to stop a car. (4) Simple to exit and re-enter on foot. Smart car, but the Stopping Framework ought to give more comfort and computerization for both businesses and clients. It ought to too meet the taking after prerequisites: (a) The framework ought to give numerous enlightening guides or enlightening to assist drivers discover a free stopping space. (b) The framework should provide viable capacities to form it less demanding for supervisors and chairmen to oversee the stopping space. (c) The framework ought to be able to anticipate unauthorized utilization. Agreeing to the over prerequisites, the brilliantly stopping framework ought to minimize human intercession and control to decrease labor costs and misplaced human mistakes and move forward effectiveness. Way better precision, strength, and adaptability, more prominent comfort for clients, lower working and support costs are moreover required from the stopping framework. The block diagram of the framework is shown in Fig. 1.

As soon as the car enters the range of this FSR, the outdoor parking display will turn on and ask for the user's identity. This makes it possible to repel outsiders looking

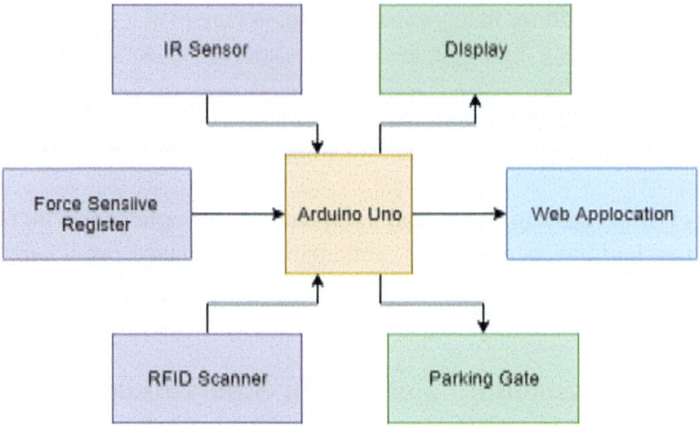

Fig. 1 Block diagram of smart car parking module

for a parking space. If the user is authorized and there is a free parking space, the parking space will be displayed on the screen. And the entrance to the parking lot opens. The user can also receive directions to the reserved parking space from the site.

Components Required:

- Arduino Uno
- FSR Sensor.

A constrain sense resistor is basically a weight sensor with a square, 1.75 × 1.5-inch identifying run. This FSR changes its resistance depending on how much weight is associated to the unstable extent. The more grounded the drive, the lower the resistance. When the FSR isn't crushed, its resistance is more critical than 1 M. This FSR of our show can recognize the associated compel between 100 g and 10 kg. As before long as the car comes beneath this FSR, the open-air stopping show will turn on and inquire for the user's personality. This permits to battle against unauthorized individuals looking to stop parcel. If the user is authorized and there is a free parking space, the parking space will be displayed on the screen. And the entrance to the parking lot opens. The user can also get directions to the reserved parking lot from the site.

Features

- Actuation Force: 0.1 N
- Force Sensitivity Range: 0.1–10.02 N
- Non-Actuated Resistance: 10 MW
- Device Rise Time: <3 μs
- Temp Operating Range: 30 to +70 °C
- 16 MHz Clock Speed
- 32 KB Flash Memory.

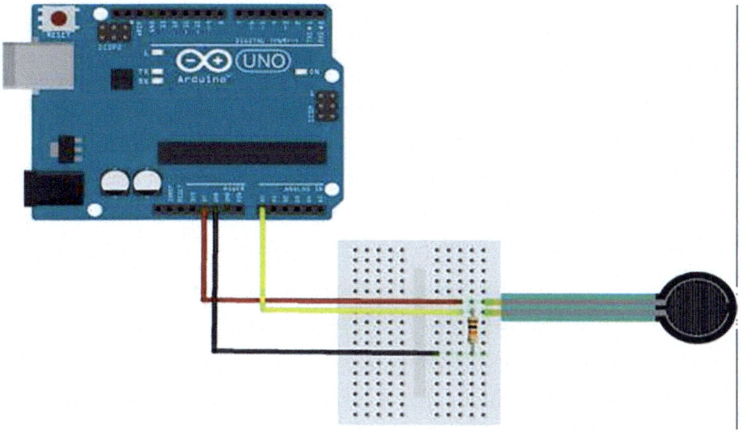

Fig. 2 FSR connection with arduino

The hardware connection between the Arduino Uno and the FSR sensor module is shown in Fig. 2. A potential divider circuit with a potentiometer is used to adjust the preset threshold of the switch D0. If the FSR reading is low, below the threshold level, the analog pin output is high. If it has a car, the pressure is high, and the output of the analog pin is low. We get the sensor output value from the analog A0 pin of the Arduino.

Applications

- Identify and Refine Press—Identify on the off chance that the touch is inadvertent or with deliberateness education. Apply constrain to client interface criticism—identify more or less client drive to form a more natural client interface.
- Make strides device security—Partitioned handle from touch for security.
- Find the center of constrain—Utilize different sensors to decide the center of drive.
- Identify nearness, area, or development—an individual or understanding in a bed, chair, or restorative gadget.
- Liquid blockage discovery—Distinguish a blockage or blockage in a pipe or pump by measuring back weight.
- Identify the correct position of the pipe.
- Many other force measurement applications.

5 RFID-Based Parking Entrance System

Radio frequency identification (RFID) employments electromagnetic areas to naturally distinguish and track markers joined to objects. Markers contain electronically put away information. Inactive markers collect vitality from examining radio swells from a close RFID collection. Dynamic markers have a unique control source comparative to a battery and can work hundreds of measures down from the RFID collection. Not at all like a standardized tag, the name doesn't have to be within the anthology's line of locate, so it can be had relations within the following thing. RFID is one framework of Automatic Identification and Data Collection (AIDC). RFID markers are utilized in various constancy, for outline, an RFID name connected to an auto amid item can be utilized to track its advance on a get-together line. RFID-labeled medications can be followed through stockpiles, and embedding RFID microchips in monster and fives permits creatures to be gratefully connected. Radio frequency identification (RFID) is the utilization of radio swells to examine and store data put away on a name connected to an object. The name can be perused from a remove of over to a few measures and doesn't have to be in coordinate line of locate of the followed collection (7–8). Components required for implementing the RFID-based parking entrance system are given by:

- Arduino Uno
- RFID Card Reader
- LCD Display
- Servo Motor.

6 Working Principle and Circuit Diagram

RFID is brief for "radio frequency identification" and alludes to an innovation where a per-user employments radio waves to capture advanced information encoded in RFID labels or keen labels (characterized underneath). In our framework (Fig. 3), the stop entryway RFID per-user peruses the flag from each RFID card and promptly sends the information to the Arduino. The token is at that point associated to the database. The stopping door is opened for each fruitful coordinate. A welcome message will moreover show up on the screen. A fluid gem show (LCD) may be a level screen or other electronic visual show that employments the light-modulating properties of fluid gems. Fluid gems don't transmit light specifically. Stick 1 (VSS) is the ground stick and this stick must be grounded for the LCD to work appropriately. VEE and VDD are regularly provided at 5 V. Be that as it may, the VEE may have a potentiometer voltage divider arranged to alter the difference. But VDD is continuously 5 V. The RS, R/W, and E pins are numbered 4, 5, and 6. RS is utilized to choose between information and command enroll. For $R_S = 0$ the command enlist is chosen and for $R_S = 1$ the information enroll is chosen. R/W permits you to select between composing and perusing. In the event that set (R/W = 1), perusing is empowered. R/W = 0 when composing. Here it is continuously based. The LCD employments the Empower pins to bolt the data shown on the information pins. 7–14 are 8-bit information pins D0–D7, which are utilized to send information to the LCD or studied the substance of the LCD's inner enroll. 15 is known as anode which is associated to Vcc. 16 is known as the cathode, 16 ought to be grounded with a 330 Ω resistor.

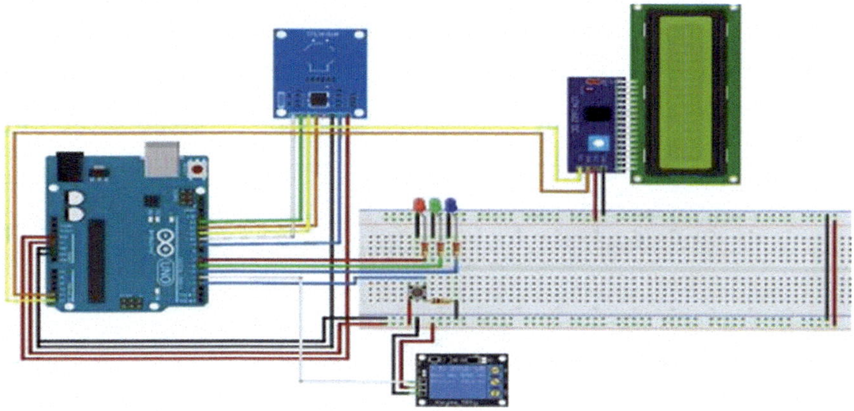

Fig. 3 RFID, motor and display connection with arduino

LCD screen integration incredibly facilitates the interaction of your advancement venture by permitting the client to straightforwardly studied a few of the yield parameters. These values can be either plain content or numerical values perused from sensors such as temperature or weight, or indeed the number of cycles completed by the Arduino. In any case, there's a little issue with these screens. When associated to a microcontroller (such as Arduino), these shows require nearly a part of interface pins, which take up nearly all the accessible IO, clearing out few yields to the multiprocessor for other gadgets and sensors. This issue is unraveled much obliged to the I2C transport communication. The LCD1602 has a coordinate microchip that oversees this sort of communication and at that point all input and yield information are restricted to as it were two PINs (but the control supply). I2C may be a sort of serial transport created by Philips that employments two-way lines called SDA (Serial Information Line) and SCL (Serial Clock Line). Both must be associated to drag up resistors. Working voltages are standard 5 and 3.3 V.

7 Physical Security System

The Internet of Things, or IoT, is the network of physically connected objects that is expanding quickly as more and more objects are being connected to the Internet. We are utilizing the extremely helpful IoT application of home security to develop a low-cost industrial and residential security system. By notifying the user anytime the door is opened or an unauthorized entrance occurs, the system will alert the owner. The user can take the required action after receiving the notification. The security system will make use of an Arduino Uno microcontroller for interfacing with other components, a magnetic Reed sensor for status monitoring, an alert buzzer, and a Wi-Fi module [9–13].

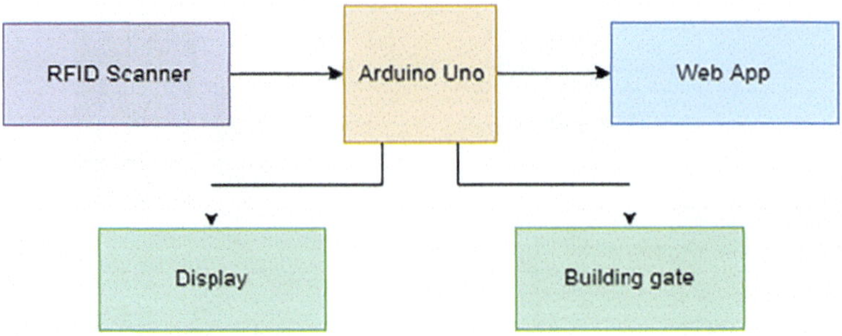

Fig. 4 Block diagram of physical security system

The provision of amenities to guarantee security, accessibility, and upkeep is the shared objective of security gates. The following considerations should be made in order to do that.

- It should be easy to utilize the system.
- To enter the building, the user needs to have a registered RFID card.

The Block Diagram of the system is illustrated in Fig. 4.

The RFID-Based Access Control System for Parking Slots is being presented here. Our system consists of a microcontroller, relay to switch the computer, LCD display, RFID reader, and power supply. An LCD display is made up of a potentiometer and one LCD. An RFID reader plus a resistor make up an RFID reader. A microcontroller called Arduino Uno is utilized. The gate is operated by a servo motor.

RFID stands for "radio-frequency identification," an innovation that employments radio waves to examine advanced information encoded in RFID labels or shrewd names (more on this afterward). Our method uses an RFID reader on the parking gate to read each RFID card's signal and send the data right away to Arduino. Next, the ID and the database are compared. Upon each successful match, the parking gate will be opened. There will also be a welcome message visible on the screen Fig. 5.

8 Smart Lighting and Connected Lighting

IoT has a significant influence on communication these days. One of its everyday applications, room automation, which is managed by an Android smartphone, was showcased in this project. The household appliances that are connected to the Arduino Uno board are managed by a Wi-Fi module (ESP8266). The Arduino Uno board's input/output ports are linked to the electrical appliances we need to operate, and serial communication is used to establish a connection between the Arduino Uno board and an Android device. This technology saves time and energy in addition to assisting in the reduction of human labor.

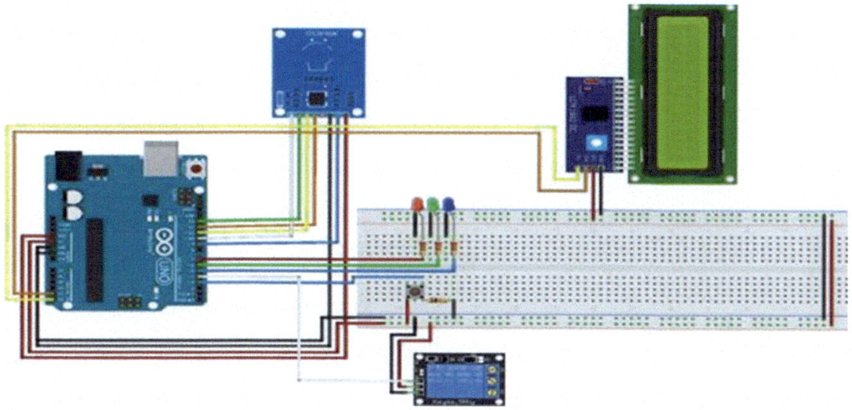

Fig. 5 RFID, motor and display connection with arduino UNO

The system's user-friendly interface, affordable construction, and simplicity of installation made it ideal for controlling household electrical appliances and devices. The ability of this automation system to automatically turn on or off the appliances based on the presence of a human in the room is its most helpful feature. This uses an infrared sensor to identify human movements. Upon entering the room, the light will automatically switch on. Additionally, after making a few indications, it will automatically turn off if there is no movement in the room. Therefore, this technique is particularly beneficial for both energy conservation and user-friendliness [7, 8].

The room lights are automatically turned on or off via a microcontroller in response to movements made by people entering and exiting the door. When developing street light management systems, automation, power consumption, and cost-effectiveness are crucial factors to take into account. The operation of the IR Sensor serves as the foundation for the project's operation. For this project, a transmissive type infrared sensor will be used. When using a transmissive infrared sensor, the transmitter and receiver are positioned in front of one another, ensuring that the detector consistently picks up infrared radiation from the transmitter. When an obstruction stands in the way of the IR Transmitter and Receiver, the IR Transmitter's rays are blocked, and the IR Receiver ceases to detect the IR Rays. Using a microcontroller, this can be set to turn the LEDs on or off. The blocked diagram of the system is shown in Fig. 6. The circuit diagrams of the system are shown in Figs. 7 and 8.

The process of autonomously operating household equipment through the use of various control system approaches is known as room automation. Numerous control strategies can be used to regulate the electrical and electronic appliances in the house, including the kitchen timer, fire alarm, lights, fans, and outdoor lighting. Apparatuses can be controlled employing an assortment of strategies, counting IoT-based domestic mechanization through the cloud, Wi-Fi-based domestic robotization by means of Android applications on any smartphone, Arduino-based domestic mechanization, farther control by means of Android applications, computerized control, RF-based

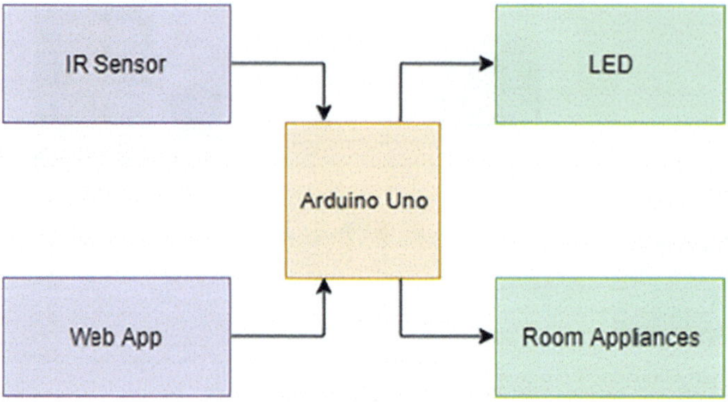

Fig. 6 Automated lighting circuit block diagram

Fig. 7 Automated lighting circuit diagram using a bread board

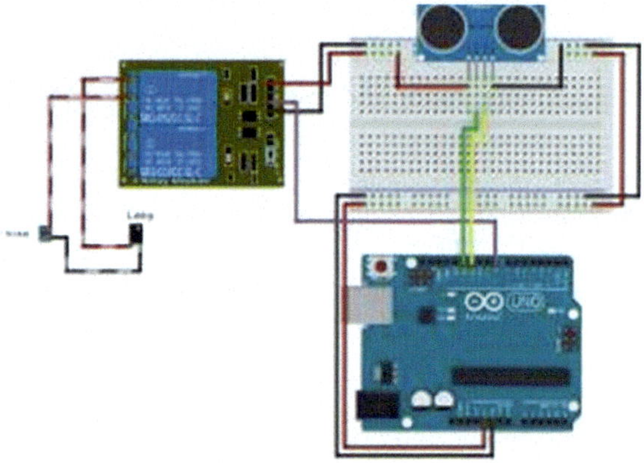

Fig. 8 IoT based appliance control circuit diagram

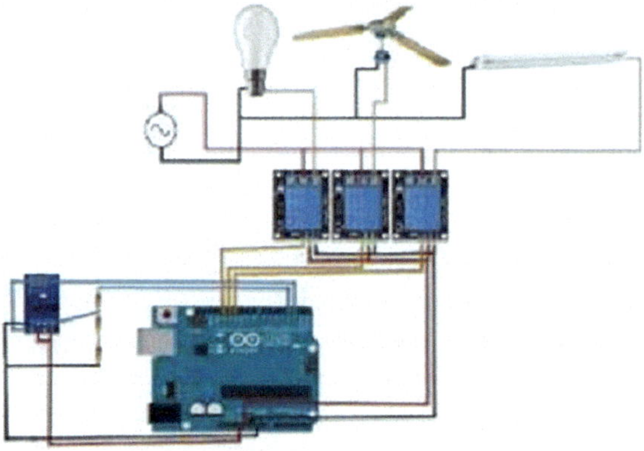

domestic robotization frameworks, and touch screen-based domestic mechanization. A web page with a front conclusion that can be customized by the client can be utilized to control and screen the stack. Through the doled-out IP address, the client can transmit commands, and these commands are bolstered Wi-Fi module. The Wi-Fi module is set up to utilize any neighborhood remote modem to get to the web. A program inside a Wi-Fi module carries out the commands that it gets.

Applications

- A variety of home automation technologies can be used to overcome the inefficiencies of standard wall switches' operation (without using conventional switching methods).
- Compared to traditional ways, home automation requires a relatively small amount of labor and can reduce electricity loss.
- Home automation systems based on IR, RF, Android applications, Arduino, Bluetooth, DTMF, etc. can operate more easily and more efficiently.
- Offers protection against electrical power short circuits when operating loads with traditional wall switches.
- Enhanced security is made possible by home automation systems that include security cameras and automated door locks.
- We can save a lot of time by employing a home automation system to control household appliances remotely (instead of having to waste time traveling from the office to the house to just unlock doors for family).

9 Photovoltaic and Renewable Energy

A sun-based shamus may be a device that orients a cargo toward the sun. Loads are by and large sun-powered boards, illustrative troughs, Fresnel mirrors, glasses or focal points. For flat-panel photovoltaic frameworks, trackers are utilized to play down the point of predominance between the approaching sun and a photovoltaic board. In recommendation, sun-powered vitality is awesome. It's clean, free, and liberal. The charge is that to specifically change over light to electrical vitality requires a sun-oriented transducer comparative to a sun-based board. The current state of this innovation clears out advance than a small to be inquired in coherence, adequacy, and taken a toll. In arrange to control a domestic or an auto requires expansive clusters of boards and undoubtedly too as it were certain geographic districts donate adequate ages of coordinate sun. Whereas sun-oriented vitality innovation isn't however to develop sufficiently to be broadly acknowledged as an essential vitality source for the common populace, it does have a few special preferences. For case, it can deliver an inaccessible vitality constraint which is extraordinary for tear charging little predisposition comparative as cell phones or GPS when traveling, climbing, or camping, or fair hanging out at the pool. It can moreover grant a source of exigency vitality for ages when the most control network is offline.

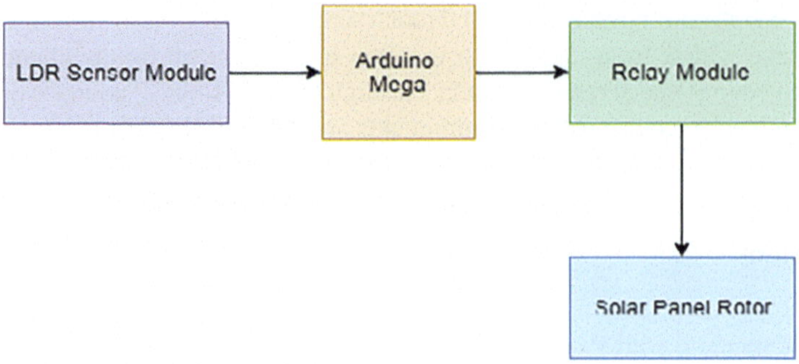

Fig. 9 Block diagram of photovoltaic and renewable energy

Solar cells used in market are fixed so it cannot get maximum sunlight throughout the day so we need a system which is able to move the solar panel according to the position of the sun. To make this module following components are required.

- Solar panel
- Arduino Uno
- LDR's X 2 (Light Dependent Resistor)
- 10 K resistors X 2
- 12 V DC gear Motor
- SPDT Relay.

The block diagram of the system is shown in Fig. 9.

Most of the times in towns or in cultivate arrive there is no power control. Hence, sun-powered vitality can be a great elective vitality source. LDR stands for Light Subordinate Resistor or Photo resistor, which may be a passive electronic component. Essentially, a resistor which contains a resistance that changes depending on the light escalated. This property permits LDR to utilize in light detecting gadget. In our venture, we are utilizing a programmed single pivot sun-based tracker. Here LDR is utilized for detecting the sun light course. The LDR voltage is detected by microcontroller and it sends signals to engine driver circuit to turn the sun-powered board to the course of sun light.

Sun-based board is associated to Stepper engine. Sun-oriented board comprises of photovoltaic cells arranged in a parallel-series arrangement. Photovoltaic cell is nothing but a sun-oriented cell. Photo takes after light and voltaic is power. Sun-oriented cell is made up of semiconductor fabric silicon. When a light beam from Sun is occurring on the sun-based cell, a few sum of vitality is retained by this fabric. The retained vitality is sufficient for the electrons to bounce from one circle to other interior of the IoTa. Cells have one or more electric field that coordinate the electrons which make current. By putting metal contact vitality can be gotten from these cells (Fig. 10).

Light Dependent Resistors are the resistors whose resistance values depend on the concentration of the light. As the escalated of light falling on the LDR increases,

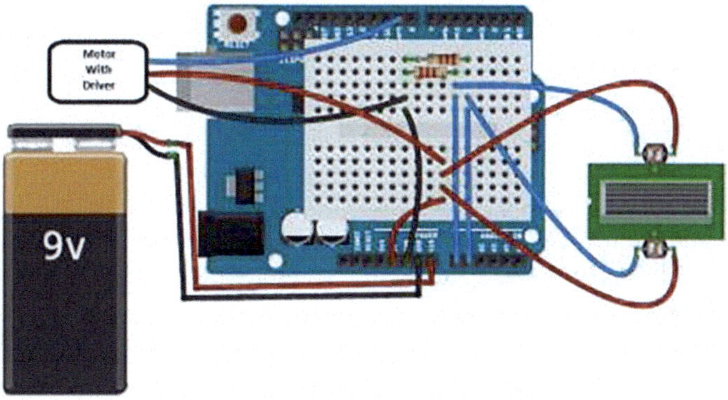

Fig. 10 Sun tracking solar panel circuit diagram

resistance esteem diminishes. In dim, LDR will have the greatest resistance. LDR will undertake an analog esteem which ought to be changed over to computerized. This will be done utilizing analog to advanced engine. ATmega16 has analog to advanced engine inside. It has six ADC channels from ADC0 to ADC5. The two LDRs are associated to ADC legs, i.e. PC0 and PC1. ADC dialog is done utilizing successive guess framework.

Step to Test the System:

- Originally power the circuit.
- Place the set up in dark.
- When the two LDRs are in dark, there's no movement in the panel.
- Now place an arsonist in front of the left LDR. Panels sluggishly move toward its left wing.
- Now move light from cleared out to right. You'll watch the board moving drowsily with the arsonist toward right.
- Within the center, when escalated on both LDRs is rise to, board.
- It'll not move until there's contrast between the light concentrated falling on the LDRs.

Advantages of Sun Tracking Solar Panel:

- The solar energy can be reused as it's nonrenewable asset.
- This moreover spares plutocrat as there's no got to pay for vitality utilized.

Limitations of Sun Tracking Solar Panel Circuit:

- In spite of the fact that sun-oriented vitality can be utilized to greatest degree this may deliver issues in stormy season.
- In spite of the fact that sun-oriented vitality can be spared to batteries,
- They're overwhelming and enthrall advance space and required to alter time to time.

10 Fire Detection and Suppression System

Fires are continuously scaring, in fact scaring. They sick individuals and destroy families with their devastation. It's basic that we upgrade fire security mindfulness and utilize firefighting to equip scholarly people to play down misfortunes and harms. Bank sensors are robotized watchers, appropriate to descry bank in realtime. However, a shrewd bank sensor will shoot data to the backend operation stage and start associated predisposition comparative as caution chimes, In case the consistency of bank surpasses a predetermined limit. The stage will too consequently shoot the data to the corresponding divisions and caution the correct labor constraint by voice dispatches or SMS. Still, conventional standalone bank sensors cannot meet the conditions of shrewd firefighting. They report false exhortations and cannot associated with individuals just like the property holders, security watches, or fire warriors. Dispersed establishment is delicate and valuable. Power-empty conventional bank sensors bear visit battery saves and are more valuable to preserve (1–13). Automated fire suppression system should be always turned on to avoid fire-related major damages.

To make prototype of the system following components are required.

- Arduino Mega.
- MQ-2 Smoke sensor.
- DHT11 temperature sensor.
- Alarm.

Block Diagram of the System is shown in Fig. 11.

Using the Internet of Things (IoT) in homes and diligence it's possible to control any electrical or electronic outfit. Also, you can get the information from any detector and assay it graphically or in any stoner-defined format from anywhere in the world. In this composition, moisture and temperature information from DHT-11 detector is analyzed graphically on Thing Speak platform using Arduino MCU.

Features

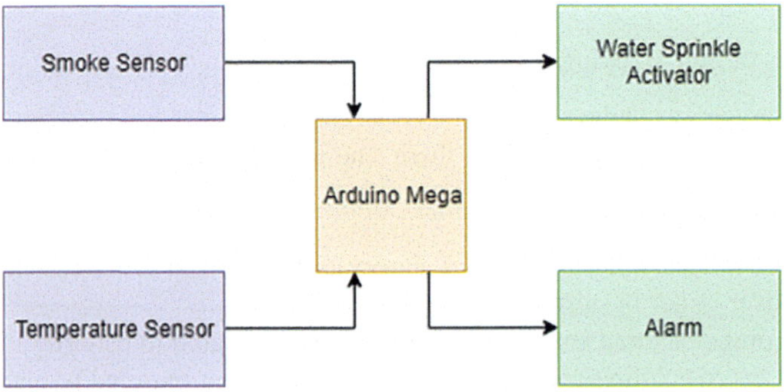

Fig. 11 Fire detection and suppression system

- Operating Voltage: 3.5–5.5 V
- Operating current: 0.3 mA (measuring) 60 μA (standby)
- Output: Serial data
- Temperature Range: 0–50 °C
- Humidity Range: 20–90%
- Resolution: Temperature and Humidity both are 16-bit
- Accuracy: ±1° C and ±1%.

11 Rooftop Gardening and Control

A building which is smart in all means needs smarter gardening ways too. Thus to save space and effectively grow plants in limited space easily, a garden can be constructed at the roof of the building. Sensors are installed to measure the soil moisture level and thus switch on a plant watering system thereby, without manual help. To make a garden automated various controls should be present in the system. In our project we are focusing on automatic watering system.

Components required to make this module are listed below.

- Arduino Mega.
- 12 V DC water pump.
- Soil moisture sensor.

Block diagram for this module is shown in Fig. 12.

Soil dampness sensors degree the volumetric water substance in soil. Since the coordinate gravimetric estimation of free soil dampness requires expelling, drying, and weighing of a test, soil dampness sensors degree the volumetric water substance by implication by utilizing a few property of the soil. The connection between the measured property and soil dampness must be calibrated and may change depending on natural components such as soil sort, temperature, or electric conductivity.

Features

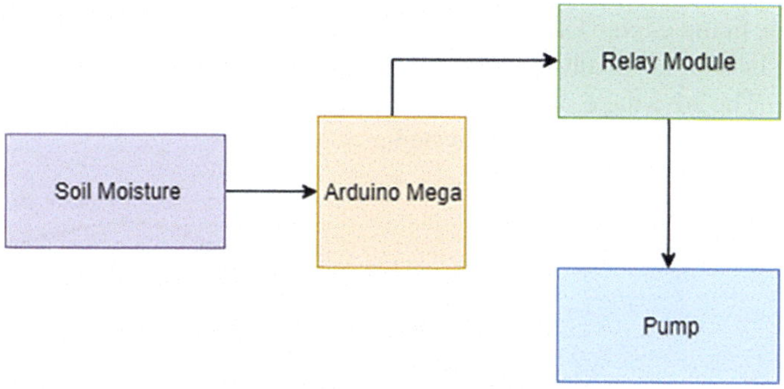

Fig. 12 Block diagram of rooftop gardening system

- Affectability movable.
- Has settled jolt entirety, helpful establishment.
- Limit level can be arranged.
- Module triple yield mode, computerized yield is basic, analog yield more exact, serial yield with correct readings.

Working Principle and Circuit Diagram:

Soil dampness sensors degree the water substance in soil. A soil dampness test is made up of different soil dampness sensors. One common sort of soil dampness sensors in commercial utilize could be a Recurrence space sensor such as a capacitance sensor. Another sensor, the neutron dampness gauge, utilizes the mediator properties of water for neutrons. Soil dampness substance may be decided through its impact on dielectric steady by measuring the capacitance between two cathodes embedded within the soil. Where soil dampness is transcendently within the frame of free water (e.g., in sandy soils), the dielectric steady is straightforwardly corresponding to the dampness substance. The test is regularly given a frequency excitation to allow estimation of the dielectric consistency. The readout from the test isn't direct with water substance and is affected by soil sort and soil temperature (Fig. 13).

As the sensor senses the water content as low, the Arduino sends a signal to the analog pin. And Arduino sends digital signal to pump. This pump is turned on until the sensor again sends a value which matches the predefined threshold value. Then the Arduino makes the pump to turn off. This way the soil moisture level of the rooftop garden is maintained.

The analog pin from soil moisture sensor will be connected to analog pin of the Arduino. Received analog value in the Arduino are mapped into a value ranging from 0 to 1023. Where lower the means higher the moisture level and higher the value equivalent to lower the moisture (Fig. 14).

Applications

- Agriculture
- Landscape irrigation.

The extension is planned to create an programmed water system framework which switches the pump engine ON/OFF on detecting the dampness substance of the soil. Within the field of horticulture, utilizing of appropriate strategy for water system is imperative. The advantage of utilizing this strategy is to diminish human mediation and still guarantee legitimate water system.

Fig. 13 Soil moisture sensor

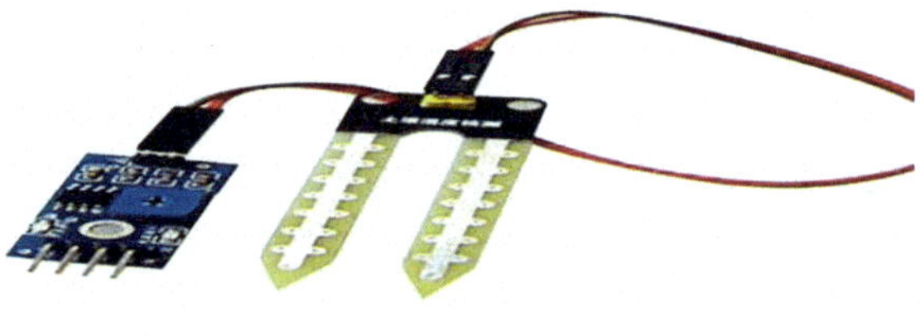

Fig. 14 Connecting soil moisture sensor to arduino

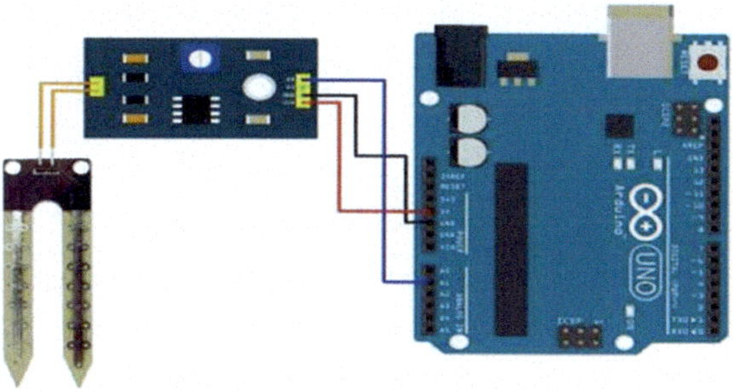

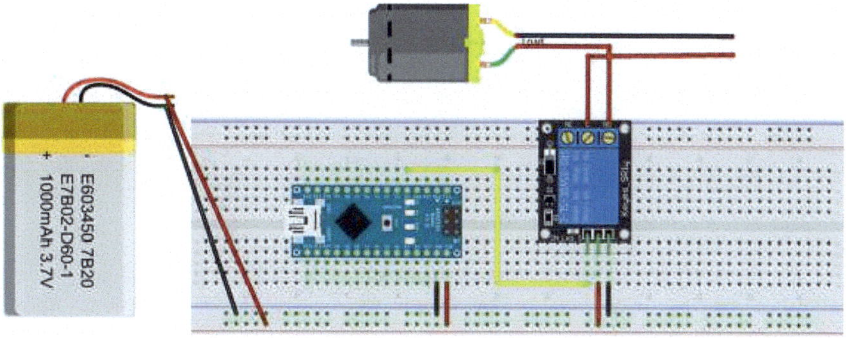

Fig. 15 Connecting pump with arduino

Pump is connected to the Arduino via a relay circuit. Pump will be on when soil moisture is low and goes off when soil moisture is high (Fig. 15).

12 Final Design Module of Smart Building Using Internet of Things (IoT)

Block Diagram of the overall system and photograph of the implemented prototype are shown in Figs. 16 and 17 respectively.

Our control board has been made of 2 development boards (Arduino UNO and Arduino MEGA). We have interfaced all the actuators, and sensors to this development boards. A wifi shield named ESp8266 has been attached to the Arduino uno board. We have created a local server using WAMP server and uploaded our database and website to that server. This system has the ability of adopting a web server also.

The final model setup of our project is shown in Fig. 17.

Now let us discuss a brief on what exactly the purpose of each module one by one.

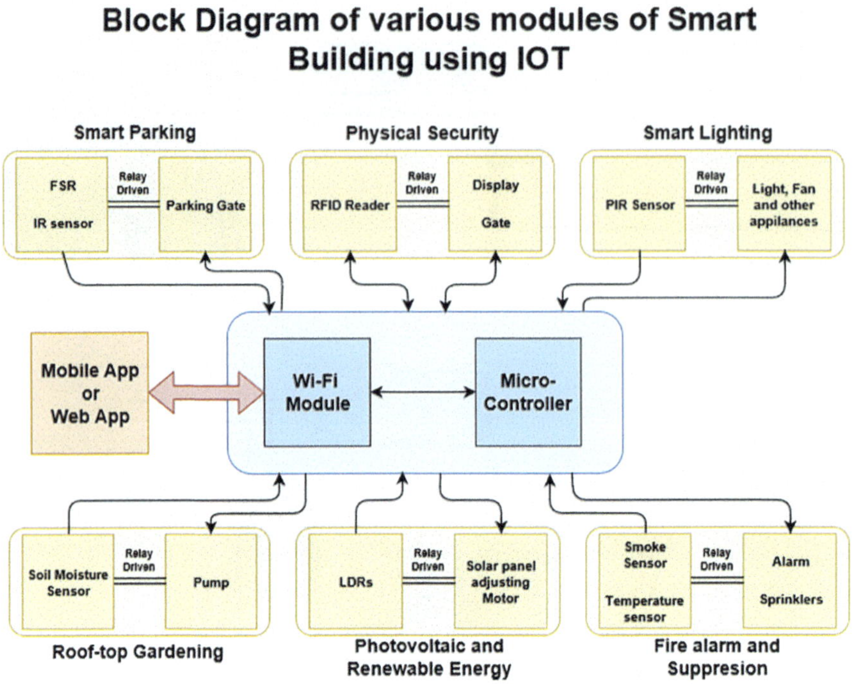

Fig. 16 Block diagram of complete system

Fig. 17 Final model display

1. Smart Parking System:
   - FSR will turn ON the setup, if it detects the presence of a car.
   - RFID Scanning is required to ACTIVATE the parking gate.
   - Available Slot no will be detected using the IR Sensor and displayed on the LCD panel on successful RFID Scanning.
2. Physical Security System:
   - RFID Scanning is required to ACTIVATE the building gate.
   - New users can be registered to the database from this system.
3. Smart Lighting and Connected Lighting:
   - IR sensor will count the presence of human being inside the room.
   - If at least one human is present inside the room the light will be turned on.
   - Light goes OFF automatically if there is no one present inside the room.
   - Room appliances can be controlled from mobile over the internet.
4. Photovoltaic and Renewable Energy:
   - Solar panel will adjust itself according to the position of the sun and it will ensure 90° angle with the sun.
   - EMF generated in the solar cell can be stored in a battery and used as an emergency backup power.
5. Fire Detection:
   - Temperature sensor and smoke sensor will detect the fire that catches inside the building.
   - It will turn on an alarm if it detects fire.
   - A water sprinkle sprayer is installed which will be activated automatically in case of fire.
6. Rooftop Gardening and Control:
   - Soil moisture sensor will measure the moisture level in the soil.
   - Pump will be automatically turned ON if soil moisture is low.
7. The Webpage:
   - Registered user can log in to the system app.
   - User can get direction to the parking slot from the app.
   - Home appliances can be controlled from the app.
   - Emergency alarm can be energized from the app.
   - User can be notified by others in the app.
   - A healthcare feature is introduced in the app for any medical emergency.

Applications

Making a smart building, or making a building smart, begins by linking core systems such as lighting, power meters, water meters, pumps, heating, fire alarms and chiller plants with sensors and control systems. At a more advanced stage, even elevators, access systems and shading can become part of the system. The main applications of the project are:

- **Smart Parking Management**:
  - Minimizes car parking complexity.
  - More organized car parking system.
- **Smart Lighting and Connected Lighting**:
  - Minimizes energy consumption.
  - Selective services can be set for individuals.
- **Photovoltaic and Renewable energy**:
  - Increases the participation in renewable energy market.
  - Can be used to light up LED bulbs in and around the building.
- **Consumer Health Services**:
  - Through IoT we can notify Emergency services for prompt action.
  - Remote monitoring products can be made compatible through the IoT network.
- **Rooftop Gardening and Control**:
  - Autonomous management of various plants is done using sensors.
  - Water pump will be automatically powered depending on the soil moisture level.

Advantages and Limitation

Advantages of our project are:

- Smart homes are convenient, they provide high end security, they are energy efficient, and they are accessible for those with special needs. Everyone wants a convenient home that makes your life easier. For some, not having to worry about things like heating and turning off lights is extremely important.
- A sense of security is crucial for all homeowners. In today's world, efficiency is essential for homeowners to both save money and save energy/power.
- Smart home technology incorporates both energy-efficient appliances and the ability to set controls so certain things only run when they need to.
- Lastly, smart home is perfect for people with special needs. The home can be programmed to do tasks that the homeowner could not do themselves.
- Smart homes too come with voice enactment innovation so somebody who couldn't ordinarily utilize a gadget would be able to by saying what they would like to do.
- The impediments of an unused innovation are fair as vital as the points of interest, in the event that not more imperative.
- People who are not computer-savvy would not want to live in a smart home since it is completely computer run. They would not be able to manage the system on their own, so they would most likely not purchase the system in the first place.
- Price is a major disadvantage since smart home technology is so expensive. Only the upper class would be able to afford a smart home system.
- Security and privacy are also main disadvantages. Since smart homes are connected through the internet, they are vulnerable to being hacked.

13 Conclusion

In spite of several features introduced for the smart building layout, there can be many smarter ways and other smart technologies that can be implemented with IoT. They are:

- Physical Security System
- Consumer Health Services.

Customers continuously search for the points of interest in modern innovations. Making or changing a building into a shrewd building is useful for both the proprietor and the organizations working inside. These benefits extend from vitality investment funds to efficiency picks up to maintainability. Solar energy has itself come up as an alternative source of effective power. Therefore utilizing this regenerative source of energy has already been initiated worldwide. This can help us in the context of this project as a generator of required power. Our main concern is to use the most of it.

IoT devices can be utilized to screen and control the mechanical, electrical, and electronic frameworks utilized in buildings (e.g., open and private, mechanical, educate, or private in domestic robotization and building mechanization frameworks. The integration of the web with building vitality administration frameworks in arrange to form vitality effective and IoT driven "Smart Building." The conceivable implies of real-time observing for diminishing vitality utilization and checking inhabitant behavior. The integration of smart devices within the built environment and how they may be utilized in future applications.

References

1. Cenedese A, Zanella A, Vangelista L, Zorzi M (2014) Padova smart city: an urban internet of things experimentation. In: Proceeding of IEEE international symposium on a world of wireless, mobile and multimedia networks 2014, Sydney, NSW, pp 1–6. https://doi.org/10.1109/WoWMoM.2014.6918931
2. Minoli D, Sohraby K, Occhiogrosso B (2017) IoT considerations, requirements, and architectures for smart buildings—energy optimization and next-generation building management systems. IEEE Internet Things J 4(1):269–283. https://doi.org/10.1109/JIOT.2017.2647881
3. https://www.intel.com/content/www/us/en/internet-of-things/smart-cities.html
4. Vlacheas P et al (2013) Enabling smart cities through a cognitive management framework for the internet of things. IEEE Commun Mag 51(6):102–111
5. Welbourne E et al (2009) Building the internet of things using RFID: the RFID ecosystem experience. IEEE Internet Comput 13(3)
6. Bridging e-health and the IoT: the SPHERE Project, The SPHERE interdisciplinary research collaboration (IRC)
7. Chen H et al (2009) The design and implementation of a smart building control system. In: IEEE international conference on e-business engineering (ICEBE'09). IEEE, pp 255–262
8. Erickson KT (1996) Programmable logic controllers. IEEE Potentials 15(1):14–17
9. Cenedese A et al (2014) Padova smart city: an urban internet of things experimentation. In: IEEE 15th international symposium on a world of wireless, mobile and multimedia networks (WoWMoM). IEEE, pp 1–6
10. Zhang D et al (2013) Efficient energy consumption and operation management in a smart building with microgrid. Energy Convers Manag 74:209–222
11. Li F, Wu C (2018) Internet of things in buildings: a review. ACM Comput Surv (CSUR) 51(5):94
12. Borgia E (2014) The Internet of things vision: key features, applications and open issues. Comput Commun 54:1–31
13. Zanella A, Bui N, Castellani A, Vangelista L, Zorzi M (2014) Internet of things for smart cities. IEEE Internet Things J 1(1):22–32

Chapter 3
Resource Management in IoT-Cloud-Based Systems

Anindita Raychaudhuri, Anwesha Mukherjee, and Debashis De

Abstract One of the key components of smart systems is Internet of Things (IoT). The IoT provides an environment where every device and sensor act as information node and data generated from these devices are transformed into useful information to the end-users. In recent days, IoT-cloud-based systems have major impact in providing necessary resources to the end-users which offers variety of applications. In spite of the potential benefit of the IoT cloud-based environment, processing of IoT data faces several challenges including resource constraints. Therefore, in a cloud-based IoT environment, resource allocation is a vital aspect. By placing the resources for computation to the network edge the latency can be decreased and the real-time processing can be improved. Numerous resource management algorithms for cloud computing-based IoT systems have been illustrated in this chapter. In this context the key challenges are identified. It also classifies the key techniques associated with resource management in IoT-cloud-based systems. Finally, we have considered two IoT-cloud-based application scenarios in the simulation, in which the first scenario has considered latency-aware online game and the second scenario has considered intelligent surveillance. Then these two application scenarios have been compared with respect to delay, network usage, and energy consumption.

Keywords Cloud computing · IoT · Radio-frequency identification (RFID) · Resource management

A. Raychaudhuri (✉)
Department of Computer Science, Sarojini Naidu College for Women, 30, Jessore Road, K.B.Sarani, Golpark, Kolkata, West Bengal 700028, India
e-mail: anindita.raychaudhuri@sncwgs.ac.in

A. Mukherjee
Department of Computer Science, Mahishadal Raj College, Garhkamalpur, Purba Medinipur, West Bengal 721628, India
e-mail: anweshamukherjee@mail.mrc.ac.in

D. De
Department of Computer Science and Engineering, Centre of Mobile Cloud Computing, Maulana Abul Kalam Azad University of Technology, NH-12, Haringhata, Nadia, West Bengal 741249, India
e-mail: debashis.de@makautwb.ac.in

A. Acharyya et al. (eds.), *Real-World Applications and Implementations of IoT*, Studies in Smart Technologies, https://doi.org/10.1007/978-981-97-8627-5_3

1 Introduction

The technological improvement in wireless network and the increasing number of users have accelerated the growth and use of smart systems in daily life. IoT serves as the key component of a smart system. Cloud computing plays a significant role in IoT systems. For providing better service, edge, and fog computing also have been integrated with IoT. Edge devices send appropriate data to the cloud only, hence, the load on the central cloud is reduced. Efficient resource allocation techniques have a crucial role in managing the relationships between the components of cloud and IoT networks, and ultimately ensure the quality of experience by the end users. Resource management is a crucial research area in IoT-cloud systems. The principal necessities and challenges of resource management in IoT are identified [1]. The authors' aim is to explore the resource management activities in IoT-cloud system. Then, the resource management in edge-cloud-based IoT systems and fog-based IoT systems are discussed. Finally, we have considered two IoT-cloud-based application scenarios in the simulation, in which the first scenario has considered online game and the second scenario has considered intelligent surveillance. Then these two application scenarios have been compared from the perspective of energy consumption, network usage, and delay. The key contribution of this chapter is that it focuses on the challenges of resource allocation, and identifies the key techniques involved in resource management in IoT-cloud systems.

Why resource management is important?

The huge data produced by sensor-equipped devices will enforce a great demand of resources for processing and storing. Resources are essential for data transformation into useful information or services. In IoT-cloud environment, resources may range from physical resources to software resources as shown in Fig. 1. Physical resource includes memory, CPU, network bandwidth, energy, etc., and software resource includes a set of programs that help to perform different tasks related to data collection, processing, storage, and evaluating instructions like to detect a complex event based on the processed data. Some applications require exhaustive processing, while other applications may be delay-sensitive. Resource management should also consider energy efficiency. Devices should be programmed to enter low-power states to reduce energy consumption. Therefore, resource management plays a vital role in an IoT-cloud environment [1–5].

What are the challenges?

There are several challenges appear in resource management for IoT systems, discussed as follows:

- Heterogeneous nature of the components used in the IoT applications.
- The implementation framework of IoT systems is highly dynamic.
- Context awareness is significant in IoT systems, that helps to build adaptive IoT systems and aids in deciding the data required to be processed on the basis of its relevance to the context provided [1].

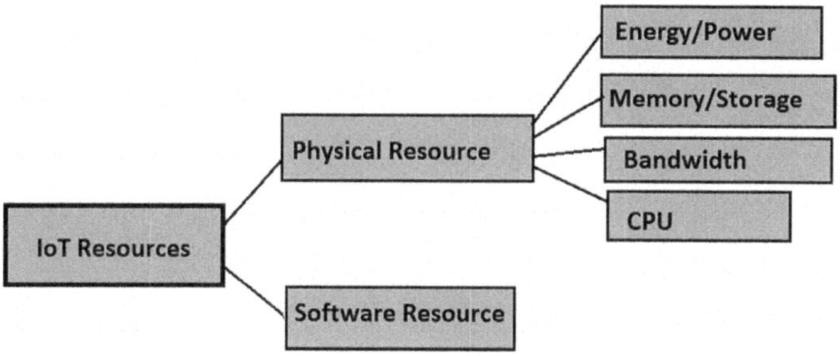

Fig. 1 Different types of resources used in IoT

- As the IoT devices are provided IP, it is possible to connect heterogeneous physical objects used in our everyday life. As the IoT devices are heterogeneous in nature, the resource allocation has become a challenge.

Layout of the chapter

Section 2 illustrates the principal resource management activities in IoT. Section 3 briefly discusses on the existing works on resource management in IoT. Section 4 presents two application scenarios in simulation and the respective results. Section 5 highlights the future scopes. The chapter is concluded in Sect. 6.

2 Principal Resource Management Activities in IoT

An IoT ecosystem consists of three tiers as described in [2], where: (i) the bottommost tier covers the IoT devices, (ii) the topmost tier includes cloud server, and (iii) the middle tier includes edge nodes or smart gateways according to the requirement of the application. The main activities involved in resource management are resource discovery, resource modeling, and resource allocation as presented in Fig. 2.

2.1 Resource Discovery

In the IoT ecosystem it is necessary to discover the essential resources before allocation. Resource discovery is itself a challenging issue especially in a heterogeneous system. Moreover, it requires to discover resources automatically in dynamic and efficient way [2]. In spite of several challenges, resource discovery [6–13] is essential as it can identify a list of authenticated resources and find services of interest to the requesting entity for smarter interactions between different IoT objects. In [14], the authors have investigated different methods for resource discovery in grid computing environment.

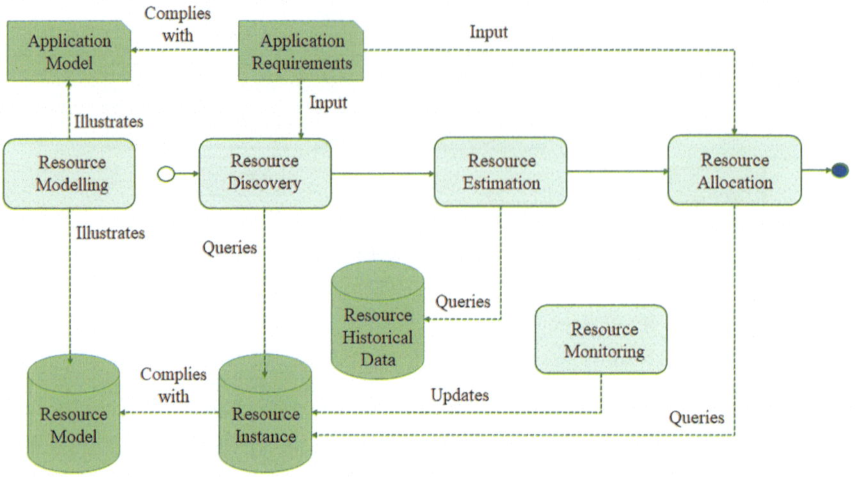

Fig. 2 Activities of resource management

2.2 Resource Modeling

Resource modeling plays a vital role in resource management, where the entities are defined along with the properties and relationships to develop the system. It should appropriately represent the components from various tiers of the IoT paradigm, starting from low-level IoT devices to high-level service interfaces.

2.3 Resource Allocation

The main objective of resource allocation is proper accommodation of the workload of all applications presently utilizing the IoT system, through allocating the essential resources. The dynamic nature of the IoT application scenario and its necessities face restrictions in time of executing parallel applications accessing the resources of the system. Therefore, resource allocation is an optimization problem that can be solved by employing different strategies.

IoT Resource Allocation Technologies: IoT can incorporate any object that is identifiable, has sensing abilities, and can communicate as needed. Machine-to-Machine (M2M) communication allows transmission between different IoT devices autonomously. The sensing data transmission to cloud server via Bluetooth or RFID which are low-power transmitters can take place for storage and analysis. Device-to-Device (D2D) communication can provide high data rates with minimal end-to-end delay because of its short-range communication. The key technologies of resource allocation are shown in Fig. 3. It has been identified that sensors, RFID tags, actuators, and mobile phones are required for the IoT applications. Edge computing is based

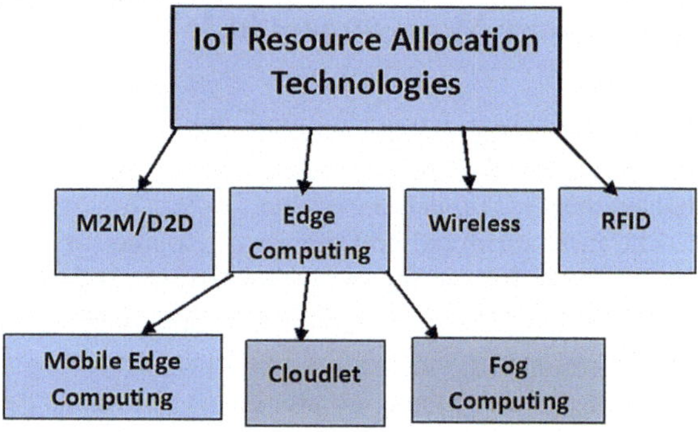

Fig. 3 Technologies associated with resource allocation in IoT

on the principle of moving resources for computation closer to things, especially to the source of the data (Table 1).

Table 1 Resource discovery mechanisms in cloud computing

Existing work	Year of publication	Key contribution	Advantages
Kalapriya et al. [6]	2004	Cloud-based mobile resource recovery	High flexibility, improved response time
Kovvur et al. [7]	2010	Resource discovery takes place using push–pull method which are adaptive in nature	Selection of unique node for job submission and adaptive resource discovery
Hasanzadeh and Meybodi [8]	2013	Use of distributed learning automata (DLA) for forwarding domain-specific queries	It optimizes resource utilization, minimizes response time, maximizes throughput, and avoids overload
Datta et al. [9]	2015	Search engine-based resource discovery framework	Supports multi-index, high scalability
Delicato et al. [10]	2017	Discovery of cloud resource specification queries	Multi-attribute dynamic queries, improved response time
Caraguay and Villalba [11]	2018	Resource discovery procedure based on the cloud computing environment	High tolerance for errors, low delay

3 Resource Management in IoT-Cloud

Many researchers have identified the significance of resource management in IoT-cloud environment [15–31]. The state-of-the-art on resource allocation and management is compared in Table 2.

The most restricted and precious resource for IoT application is battery power. Hence, various researchers have focused on minimizing the energy consumption [32–37]. Improving storage capabilities is another challenging issue that can be managed by caching storage capacities [38]. IoT devices generally suffer from restricted processing power. Managing processing power can decrease the computation delay, as well as the response time can be minimized [39, 40]. Service management plays a crucial role in optimizing scheduling service and Quality of Service (QoS) [41–43]. Throughput of the system depends on the adaptive technique of resource reservation [44] and channel management [45].

Table 3 summarizes different resource allocation approaches. The resource allocation approaches such as energy-resource allocation based on optimized framework, optimization in the placement of virtual machine in cloud environment, etc., have

Table 2 Existing works on resource management in IoT-cloud environment

Existing work	Year of publication	Focus of the study
Mohamaddiah et al. [15]	2014	Resource allocation and monitoring
Bhavani and Guruprasad [16]	2014	Resource allocation techniques
Singh and Chana [17]	2016	Resource scheduling approaches
Lavanya and Bindy [18]	2016	Resource scheduling approaches
Hameed et al. [19]	2016	Energy efficient approaches
Gu et al. [20]	2018	Resource allocation to optimize system performance in IoT-fog computing
Li and Liutong [21]	2019	Edge-cloud collaborative computational resource management strategy
Javadpour et al. [22]	2020	Resource management issues in IoT-cloud environment
Gao et al. [23]	2021	Resource allocation in industrial IoT for federated learning
Xavier et al. [24]	2022	Collaborative and energy-aware resource allocation
Li et al. [25]	2022	Deep reinforcement learning-based resource management
Cui et al. [26]	2023	Minimization of network latency
Kumar et al. [27]	2023	In fog-enabled industrial IoT, autonomous nature of workload prediction as well as resource allocation are explored

Table 3 Different approaches of resource allocation associated with each category of resource

Article	Year of publication	Resource type	Focus of the study
Jiang et al. [32]	2016	Energy	Algorithm focuses on energy-efficient routing, which also offers load balancing
Liu and Ansari [33]	2017	Energy	Architecture based on green relay assisted D2D communications
Khalil et al. [34]	2018	Energy	Protocol to minimize energy expenditure
Mishra et al. [35]	2018	Energy	Energy harvesting design
Yang et al. [36]	2017	Energy	Multiple access techniques
Zhang et al. [37]	2019	Energy	Online two time-scale resource allocation algorithms
Yao and Ansari [38]	2019	Storage	Hierarchical cache-enabled cloud-radio access network architecture to optimize the storage allocation problem
Shah-Mansouri and Wong [39]	2018	Processing power	Allocation of fog computing resources in close proximity of IoT users for better computing
Fan and Ansari [40]	2018	Processing power	Minimization of the response time
Angelakis et al. [41]	2016	Service	Service assignment using heterogeneous resources
Ding et al. [42]	2017	Service	Optimization of selection of service nodes
Aron [43]	2017	Service	Model that can host concurrent applications and handle all issues related to this while facilitating
Albasheir and Kodoch [44]	2016	Bandwidth	Adaptive method that enhances reservation of resources for long-term evolution (LTE) mobile guaranteed services
de Andrade [45]	2016	Channel	Allocation of resource to prioritize control message and control channel

been proposed by many researchers [46–55]. Sensor nodes are important components of IoT. Therefore, the energy consumption minimization in wireless sensor network has a major impact in energy management of IoT-cloud network. The utilization of bio-inspired approaches can have a major impact in this case [56–64].

4 Simulation Results and Discussion

In this section, we consider two application scenarios of IoT system. In the first scenario, we have considered delay-aware online game [65]. In the second scenario, we have considered intelligent surveillance [65]. For simulation iFogSim simulator has been used.

4.1 Delay-Aware Online Game

In this case, electroencephalography (EEG) Tractor Beam Game that contains Client, Concentration calculator, and Coordinator, is considered [65]. The CPU of the Internet Service Provider (ISP) gateway, smartphone, Wi-Fi gateway, and cloud virtual machine (VM) are considered 3 GHz, 1.6 GHz, 3 GHz, and 3 GHz respectively. The RAM of ISP gateway, smartphone, Wi-Fi gateway, and cloud VM is considered 4 GB, 1 GB, 4 GB, and 4 GB respectively. The power consumption of ISP gateway, Wi-Fi gateway, and cloud VM in active state and idle state are considered 107.339 W and 83.433 W respectively. The power consumption of the smartphone in active state and idle state are considered 87.53 W and 82.44 W respectively. The delay from EEG headset to smartphone is 6 milliseconds (ms). The delay from smartphone to Wi-Fi gateway is 2 ms. The delay from Wi-Fi gateway to ISP gateway is 4 ms. The delay from ISP gateway to cloud data center is 100 ms. The tuple CPU length of two headsets A and B are considered 2000 million instructions and 2500 million instructions respectively. The average inter-arrival time of A and B is considered 10 ms and 5 ms respectively.

We have considered five case scenarios depending on the number of Wi-Fi gateways, where each gateway is connected to four smartphones. The number of Wi-Fi gateways in case scenarios 1, 2, 3, 4, and 5 are 1, 2, 4, 8, and 16 respectively. Figure 4 shows the energy consumption and network usage for the considered application. The average delay for edge-ward placement and cloud-only model are presented in Table 4. We observe that the delay for edge-ward placement is 100 ms for the five case scenarios. For the cloud-only model the delay is 250 ms, 260 ms, 270 ms, 800 ms, and 4000 ms for case scenario 1, 2, 3, 4, and 5 respectively for headset A. For headset B, the delay for the cloud-only model is 250 ms, 260 ms, 270 ms, 2900 ms, and 4550 ms for case scenario 1, 2, 3, 4, and 5 respectively. This is observed that the delay in edge-ward placement is ~78% lower than the system where only cloud resources are used.

The energy consumptions of the devices in fog-based and cloud-only models are presented in Fig. 4a. This is observed that the use of fog devices reduces cloud data centers' energy consumption while increasing the edge devices' energy consumption slightly. We observe that the network usage of the two headsets in edge-ward placement is ~77% less than the cloud-only model (refer to Fig. 4b). The network usage is measured in kilobytes (KB). With the increase in the connected number of devices,

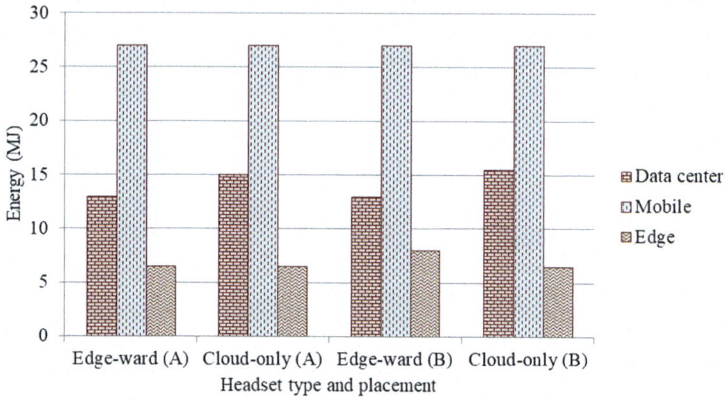

(a) Energy consumption

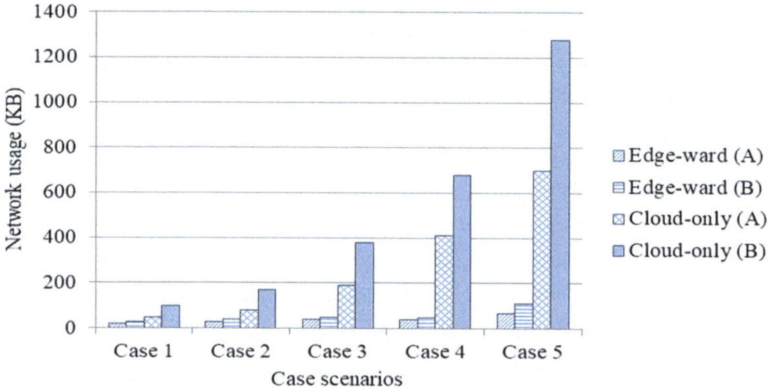

(b) Network usage

Fig. 4 Energy consumption and network usage for application scenario 1

Table 4 Average delay in edge-ward placement and cloud-only model for application scenario 1

Case scenario	Average delay (ms)			
	Edge-ward (headset A)	Edge-ward (headset B)	Cloud-only (headset A)	Cloud-only (headset B)
Case 1	100	100	250	250
Case 2	100	100	260	260
Case 3	100	100	270	270
Case 4	100	100	800	2900
Case 5	100	100	4000	4550

the network load is also increased if only cloud resources are used. An uncontrolled growth of network usage should be avoided because it can cause network congestion and performance degradation. Hence, the use of only cloud-based system should be avoided.

4.2 DCN-Based Intelligent Surveillance

In this case, an intelligent surveillance application using Distributed camera networks (DCN) has been considered [65]. The application contains object tracker, object detector, motion detector, Pan-Tilt zoom (PTZ) control, and a user interface [65]. The CPU of the area gateway and ISP gateway is considered 3 GHz. The RAM of ISP gateway and area gateway is considered 4 GB. The power consumption of area gateway, ISP gateway, and cloud VM in active state and idle state are considered 107.339 W and 83.433 W respectively. The delay from camera to area switch is 2 ms. The delay from area gateway and ISP gateway is 2 ms. The delay from ISP gateway to cloud data center is 100 ms. The CPU length, network length, and average inter-arrival time for the sensor for the surveillance application are considered 1000 million instructions, 2000 bytes, and 5 ms respectively.

Three case scenarios have been considered depending on the number of areas under surveillance. The number of areas under surveillance in case scenarios 1, 2, and 3 are 2, 4, and 8 respectively. The energy consumption and network usage for the considered application are presented in Fig. 5. The average delay for edge-ward placement and cloud-only model are presented in Table 5. We observe that the delay for edge-ward placement is 8.5 ms for the three case scenarios. For the cloud-only model the delay is 210.8 ms, 211.6 ms, and 1293.9 ms for case scenarios 1, 2, and 3 respectively. This is observed that the delay in edge-ward placement is ~97% lower than the system where only cloud resources are used.

This is observed that the delay in edge-ward placement is much lower than the system where only cloud resources are used. The energy consumptions of the devices in fog-based and cloud-only models are presented in Fig. 5b. As the cameras perform motion detection, the energy consumption is increased. This is observed that use of fog devices reduces the cloud data centers' energy consumption. We observe that the network usage of the edge-ward placement is ~91% less than the cloud-only model (refer to Fig. 5a). As in edge-ward placement, fog-based execution takes place, and modules such as object tracker, object detector, are placed on the edge devices, as a result the data volume sent to the cloud data center is reduced.

From the two application scenarios of IoT systems, we observe that using edge/ fog computing the delay can be reduced than the cloud-only systems.

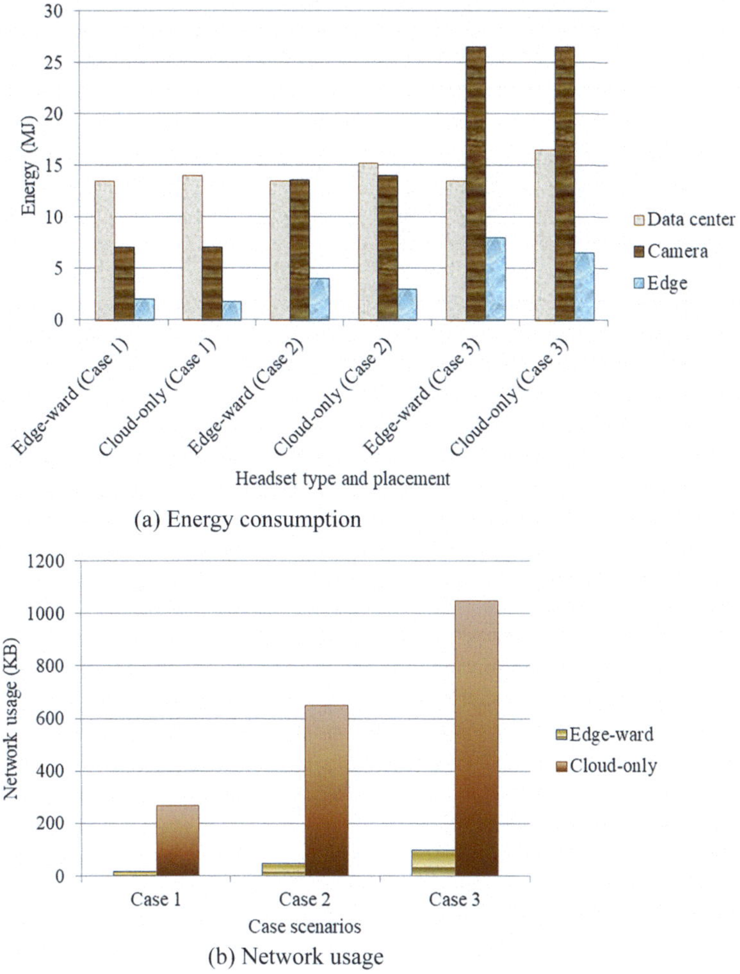

Fig. 5 Energy consumption and network usage for application scenario 2

Table 5 Average delay in edge-ward placement and cloud-only model for application scenario 2

Case scenario	Average delay (ms)	
	Edge-ward placement	Cloud-only model
Case 1	8.5	210.8
Case 2	8.5	211.6
Case 3	8.5	1293.9

5 Future Research Scopes

We have discussed on the existing resource management methods for IoT systems in the previous sections. Though, several researches on resource management for IoT exist, various research scopes are present for cloud computing-based IoT systems, such as [66]:

- Energy-aware resource management
- Machine learning-based optimal resource allocation
- Delay-aware resource management
- Cost-effective resource management
- Reliable resource management
- Blockchain-based resource management.

Minimization of energy consumption is one of the major objectives. Thus, energy-efficient resource allocation and management is a significant research direction. The vast amount of data collected by the IoT devices are required to process. For decision-making, machine learning has a major impact. For big data analysis, the use of deep learning is also important. Though, by using edge and fog computing the delay has been reduced compared to the cloud-only frameworks, for real-time applications delay minimization is still an important issue especially for hard deadline tasks. Cost-effective resource management is another significant aspect. Reliability is another major concern. The use of federated learning [67] for dealing with this issue is another future scope. For data security and privacy preservation, blockchain also plays a significant role [66].

6 Conclusion

In this chapter, resource management in IoT systems has been discussed. The principal resource management activities, the type of resources, and the existing methods on resource management for IoT systems are discussed. We have illustrated various resource management algorithms for cloud computing-based IoT systems along with the edge and fog computing-based IoT systems. The simulation of IoT system based on cloud computing has been performed considering two different types of applications. The simulation results present that by using edge and fog computing the delay can be reduced than the cloud-only systems. Finally, the future research directions in IoT resource management are highlighted.

References

1. Perera C, Zaslavsky A, Christen P, Georgakopoulos D (2013) Context aware computing for the internet of things: a survey. IEEE Commun Surv Tutor 16(1):414–454
2. Delicato FC, Pires PF, Batista T (2017) The resource management challenge in IoT. In: Resource management for internet of things. SpringerBriefs in computer science. Springer, Cham
3. Sadri AA, Rahmani AM, Saberikamarposhti M, Hosseinzadeh M (2022) Data reduction in fog computing and internet of things: a systematic literature survey. Internet Things 100629
4. Maray M, Shuja J (2022) Computation offloading in mobile cloud computing and mobile edge computing: survey, taxonomy, and open issues. Mob Inf Syst 2022
5. Jain V, Kumar B, Gupta A (2022) Cybertwin-driven resource allocation using deep reinforcement learning in 6G-enabled edge environment. J King Saud Univ Comput Inf Sci 34(8):5708–5720
6. Kalapriya K, Nandy SK, Srinivasan D, Uma Maheshwari R, Satish V (2004) A framework for resource discovery in pervasive computing for mobile aware task execution. In: Proceedings of the 1st conference on computing frontiers, pp 70–77
7. Kovvur RMR, Kadappa V, Ramachandram S, Govardhan A (2010) Adaptive resource discovery models and resource selection in grids. In: 2010 First international conference on parallel, distributed and grid computing (PDGC 2010), pp 95–100
8. Hasanzadeh M, Meybodi MR (2013) Grid resource discovery based on distributed learning automata. Computing 96:909–922
9. Datta SK, Da Costa RP, Bonnet C (2015) Resource discovery in internet of things: current trends and future standardization aspects. In: 2015 IEEE 2nd world forum on internet of things (WF-IOT), pp 542–547
10. Delicato FC, Pires PF, Batista T (2017) The activities of resource discovery and resource estimation. In: Resource management for internet of things. SpringerBriefs in computer science. Springer, Cham, pp 33–44
11. Caraguay ÁLV, Villalba LJG (2017) Monitoring and discovery for self-organized network management in virtualized and software defined networks. Sensors 17(4):731
12. Khalil K, Khalid E, Mohamed S, Magdy B (2020) Resource discovery techniques in the internet of things: a review. Internet Things 12:100293
13. Liu W, Nishio T, Shinkuma R, Takahashi T (2014) Adaptive resource discovery in mobile cloud computing. Comput Commun 50:119–129
14. Navimipour NJ, Rahmani AM, Navin AH, Hosseinzadeh M (2014) Resource discovery mechanisms in grid systems: a survey. J Netw Comput Appl 41:389–410
15. Mohamaddiah MH, Abdullah A, Subramaniam S, Hussin M (2014) A survey on resource allocation and monitoring in cloud computing. Int J Mach Learn Comput 4(1):31–38
16. Bhavani BH, Guruprasad HS (2014) A comparative study on resource allocation policies in cloud computing environment. Compusoft 3(6):893
17. Singh S, Chana I (2016) A survey on resource scheduling in cloud computing: issues and challenges. J Grid Comput 14:217–264
18. Lavanya BM, Bindu CS (2016) Systematic literature review on resource allocation and resource scheduling in cloud computing. Int J Adv Inf Technol (IJAIT) 6(4):1–15
19. Hameed A, Khoshkbarforoushha A, Ranjan R, Jayaraman PP, Kolodziej J, Balaji P, Zeadally S, Malluhi QM, Tziritas N, Vishnu A, Khan SU, Zomaya A (2016) A survey and taxonomy on energy efficient resource allocation techniques for cloud computing systems. Computing 98:751–774
20. Gu Y, Chang Z, Pan M, Song L, Han Z (2018) Joint radio and computational resource allocation in IoT fog computing. IEEE Trans Veh Technol 67(8):7475–7484
21. Li Y, Xu L (2019) The service computational resource management strategy based on edge-cloud collaboration. In: 2019 IEEE 10th international conference on software engineering and service science (ICSESS). IEEE, pp 400–404
22. Javadpour A, Wang G, Rezaei S (2020) Resource management in a peer to peer cloud network for IoT. Wirel Pers Commun 115:2471–2488

23. Gao W, Zhao Z, Min G, Ni Q, Jiang Y (2021) Resource allocation for latency-aware federated learning in industrial internet of things. IEEE Trans Industr Inf 17(12):8505–8513
24. Xavier TC, Delicato FC, Pires PF, Amorim CL, Li W, Zomaya A (2022) Managing heterogeneous and time-sensitive IoT applications through collaborative and energy-aware resource allocation. ACM Trans Internet Things 3(2):1–28
25. Li Q, Kumar P, Alazab M (2022) IoT-assisted physical education training network virtualization and resource management using a deep reinforcement learning system. Complex Intell Syst 1–14
26. Cui T, Yang R, Fang C, Yu S (2023) Deep reinforcement learning-based resource allocation for content distribution in iot-edge-cloud computing environments. Symmetry 15(1):217
27. Kumar M, Kishor A, Samariya JK, Zomaya AY (2023) An autonomic workload prediction and resource allocation framework for fog enabled industrial IoT. IEEE Internet Things J
28. Sedaghat M, Hernández-Rodríguez F, Elmroth E (2014) Autonomic resource allocation for cloud data centers: a peer to peer approach. In: 2014 International conference on cloud and autonomic computing, pp 131–140
29. Javadpour A (2020) Providing a way to create balance between reliability and delays in SDN networks by using the appropriate placement of controllers. Wirel Pers Commun 110:1057–1071
30. Javadpour A (2019) Improving resources management in network virtualization by utilizing a software-based network. Wirel Pers Commun 106:505–519
31. Li X, Da Xu L (2020) A review of internet of things—resource allocation. IEEE Internet Things J 8(11):8657-8666
32. Jiang D, Zhang P, Lv Z, Song H (2016) Energy-efficient multi-constraint routing algorithm with load balancing for smart city applications. IEEE Internet Things J 3(6):1437–1447
33. Liu X, Ansari N (2017) Green relay assisted D2D communications with dual batteries in heterogeneous cellular networks for IoT. IEEE Internet Things J 4(5):1707–1715
34. Khalil EA, Ozdemir S, Tosun S (2018) Evolutionary task allocation in internet of things-based application domains. Futur Gener Comput Syst 86:121–133
35. Mishra D, Alexandropoulos GC, De S (2018) Energy sustainable IoT with individual QoS constraints through MISO SWIPT multicasting. IEEE Internet Things J 5(4):2856–2867
36. Yang Z, Xu W, Pan Y, Pan C, Chen M (2017) Energy efficient resource allocation in machine-to-machine communications with multiple access and energy harvesting for IoT. IEEE Internet Things J 5(1):229–245
37. Zhang D, Qiao Y, She L, Shen R, Ren J, Zhang Y (2018) Two time-scale resource management for green Internet of Things networks. IEEE Internet Things J 6(1):545–556
38. Yao J, Ansari N (2019) Joint content placement and storage allocation in C-RANs for IoT sensing service. IEEE Internet Things J 6(1):1060–1067
39. Shah-Mansouri H, Wong VWS (2018) Hierarchical fog-cloud computing for IoT systems: a computation offloading game. IEEE Internet Things J 5(4):3246–3257
40. Fan Q, Ansari N (2018) Application aware workload allocation for edge computing-based IoT. IEEE Internet Things J 5(3):2146–2153
41. Angelakis V, Avgouleas I, Pappas N, Fitzgerald E, Yuan D (2016) Allocation of heterogeneous resources of an IoT device to flexible services. IEEE Internet Things J 3(5):691–700
42. Ding S, He X, Wang J (2017) Multiobjective optimization model for service node selection based on a tradeoff between quality of service and resource consumption in mobile crowd sensing. IEEE Internet Things 4(1):258–268
43. Aron R (2017) IoT based platform as a service for provisioning of concurrent applications. arXiv:1711.10685
44. Albasheir S, Kodoch M (2016) Enhanced control for adaptive resource reservation of guaranteed services in LTE networks. IEEE Internet Things J 3(2):179–189
45. de Andrade TP, Astudillo CA, da Fonseca NL (2016) Allocation of control resources for machine-to-machine and human-to-human communications over LTE/LTE-A networks. IEEE Internet Things J 3(3):366–377

46. Alam S, De D, Ray A (2015 May) Analysis of energy consumption for IARP, RIP and STAR routing protocols in wireless sensor networks. In: 2015 Second international conference on advances in computing and communication engineering, pp 11–16
47. Raychaudhuri A, Mukherjee A, De D (2021) SMEC: sensor mobile edge computing. Mob Edge Comput 89–110
48. Jha RS, Gupta P (2016) Power & load aware resource allocation policy for hybrid cloud. Procedia Comput Sci 78:350–357
49. Al-Fuqaha A, Guizani M, Mohammadi M, Aledhari M, Ayyash M (2015) Internet of things: a survey on enabling technologies, protocols, and applications. IEEE Commun Surv Tutor 17(4):2347–2376
50. Jayaprakash S, Nagarajan MD, Prado RPD, Subramanian S, Divakarachari PB (2021) A systematic review of energy management strategies for resource allocation in the cloud: clustering, optimization and machine learning. Energies 14(17):5322
51. Goyal S, Bhushan S, Kumar Y, Rana AUHS, Bhutta MR, Ijaz MF, Son Y (2021) An optimized framework for energy-resource allocation in a cloud environment based on the whale optimization algorithm. Sensors 21(5):1583
52. Azizi S, Zandsalimi MH, Li D (2020) An energy-efficient algorithm for virtual machine placement optimization in cloud data centers. Clust Comput 23:3421–3434
53. Zahedi Fard SY, Ahmadi MR, Adabi S (2017) A dynamic VM consolidation technique for QoS and energy consumption in cloud environment. J Supercomput 73(10):4347–4368
54. Armbrust M, Fox A, Griffith R, Joseph AD, Katz R, Konwinski A, Lee G, Patterson D, Rabkin A, Stoica I, Zaharia M (2010) A view of cloud computing. Commun ACM 53(4):50–58
55. Manna S, Bhunia SS, Mukherjee N (2014) Vehicular pollution monitoring using IoT. In: International conference recent advanced innovations engineering, pp 1–5
56. Raychaudhuri A, De D (2020) Bio-inspired algorithm for multi-objective optimization in wireless sensor network. In: De D, Mukherjee A, Kumar Das S, Dey N (eds) Nature inspired computing for wireless sensor networks. Springer tracts in nature-inspired computing, Springer, Singapore
57. Raychaudhuri A, Mukherjee A, De D, Gill SS (2022) Green Internet of Things using mobile cloud computing: architecture, applications, and future directions. Green mobile cloud computing. Springer International Publishing, Cham, pp 213–229
58. Ray A, De D (2016) Energy efficient clustering protocol based on K-means (EECPK-means)-midpoint algorithm for enhanced network lifetime in wireless sensor network. IET Wirel Sens Syst 6(6):181–191
59. Bakshi M, Ray A, De D (2017) Maximizing lifetime and coverage for minimum energy wireless sensor network using corona based sensor deployment. CSI Trans ICT 5:17–25
60. Dashti SE, Rahmani AM (2016) Dynamic VMs placement for energy efficiency by PSO in cloud computing. J Exp Theor Artif Intell 28:97–112
61. Gao Y, Guan H, Qi Z, Hou Y, Liu L (2013) A multi-objective ant colony system algorithm for virtual machine placement in cloud computing. J Comput Syst Sci 79(8):1230–1242
62. Kansal NJ, Chana I (2015) Artificial bee colony based energy-aware resource utilization technique for cloud computing. Concurr Comput Pract Exp 27(5):1207–1225
63. Raychaudhuri A, De D (2020) Bio-inspired algorithm for multi-objective optimization in wireless sensor network. In: Nature inspired computing for wireless sensor networks, pp 279–301
64. Alroobaea R, Arul R, Rubaiee S, Alharithi FS, Tariq U, Fan X (2022) AI-assisted bio-inspired algorithm for secure IoT communication networks. Clust Comput 25(3):1805–1816
65. Gupta H, Vahid Dastjerdi A, Ghosh SK, Buyya R (2017) iFogSim: a toolkit for modeling and simulation of resource management techniques in the internet of things, edge and fog computing environments. Softw Pract Exp 47(9):1275–1296
66. Ali SA, Ansari M, Alam M (2020) Resource management techniques for cloud-based IoT environment. In: Internet of things (IoT) concepts and applications, pp 63–87
67. Wang L, Zhao X, Lu Z, Wang L, Zhang S (2023) Enhancing privacy preservation and trustworthiness for decentralized federated learning. Inf Sci 628:449–468

Chapter 4
Smart Street Lighting Systems: A Revolutionary IoT Solution for Large-Scale Electrical Energy Conservation

Aritra Acharyya

Abstract This chapter introduces an innovative approach designed to enhance the street lighting infrastructure in city/town. This ground-breaking method combines a long-distance Internet Service Provider (ISP) sharing strategy, offering an efficient solution for the deployment of a comprehensive smart lighting network across urban and suburban regions, and potentially highways and railway tracks. A significant advantage of this approach lies in its substantially lower installation and maintenance costs compared to existing LoRaWAN-based smart street lighting systems. The smart street lights featured in this chapter offer a wide range of essential functionalities, including automated illumination, dynamic dimming, remote control, and real-time monitoring. Furthermore, the plan includes the establishment of long-distance ISP sharing wireless local area networks (WLANs) strategically covering critical streets in a city or town. These networks serve as the foundation for the wireless Internet Protocol (IP) cameras at strategic locations, enabling real-time traffic monitoring and remote surveillance. In summary, this technology not only offers a cost-effective solution but also incorporates advanced technology to elevate the town's infrastructure and enhance safety measures, making it a noteworthy contribution to the field of smart street lighting systems and large-scale electrical energy conservation.

Keywords Smart city · Internet-of-Things (IoT) · Smart street lights · LED lights · WiFi · Wireless communication · 2.4 GHz

---

A. Acharyya (✉)
Department of Electronics and Communication Engineering, Kalyani Government Engineering College, Kalyani, Nadia, West Bengal 741235, India
e-mail: ari_besu@yahoo.co.in

A. Acharyya et al. (eds.), *Real-World Applications and Implementations of IoT*, Studies in Smart Technologies, https://doi.org/10.1007/978-981-97-8627-5_4

1 Introduction

An essential component for electrical power consumption reduction in smart cities is the Internet-of-Things (IoT) based smart street lighting system. Currently, this smart street lighting system is widely prevalent in the United States of America and European nations [1, 2]. Among the array of smart cities, Ahmedabad in Gujarat stands out as a pioneer, having implemented its smart street lighting system over the past decade [3]. The city has successfully deployed over two hundred thousand smart LED street lights to date. Key attributes of these smart street lights include:

(i) *Automated Illumination*: The LED street lights are designed to automatically switch on at dusk and off at dawn by sensing ambient sunlight levels.
(ii) *Dynamic Dimming*: These intelligent LED street lights possess dimmable capabilities. During late-night hours, when there is no vehicular or pedestrian activity, the lights intelligently dim to conserve energy, often reducing power consumption by half or more. However, they swiftly return to full brightness upon detecting motion in their vicinity.
(iii) *Remote Control*: The system allows for remote control operations from a central control room. This enables the lights to be turned on, turned off, or adjusted in brightness level, all managed conveniently from a centralized location.
(iv) *Real-time Monitoring*: A graphical user interface (GUI) provides the ability to remotely monitor the operational status of each light from the control room. This feature aids in promptly identifying malfunctioning street lights, facilitating swift replacements by the relevant authorities when needed.

The IoT-based smart street lighting system proves to be a pivotal tool in enhancing energy efficiency and maintenance efficacy within urban environments. The Smart City Ahmedabad Development Limited is employing an advanced technology framework to establish a comprehensive global smart city. This framework encompasses a LoRa Wide Area Network (LoRaWAN) and an IoT-based smart street lighting system, in addition to transit management, city surveillance, e-governance, and an integrated command and control center. However, the adoption of the LoRaWAN technology presents certain challenges due to its high cost and recent introduction. Consequently, the overall cost-effectiveness of the smart street lighting system could be hindered by substantial installation and maintenance expenses. Furthermore, the utilization of a relatively less mature technology like LoRaWAN introduces complexities in maintaining and advancing the smart street lighting system over time.

This chapter introduces an innovative approach for enhancing the lighting system in Cooch Behar town. The suggested method combines a long-distance Internet Service Provider (ISP) sharing technique like metropolitan area network (MAN). This technology proves effective for implementing a comprehensive smart lighting system across both urban and suburban areas, and it can also be seamlessly extended to cover highway and railway lighting systems. Importantly, the installation and upkeep expenses associated with this proposed technology are significantly lower compared to a LoRaWAN-based smart street lighting system. Moreover, the proposed solution envisions the deployment of long-distance ISP sharing wireless local area networks (WLANs) which span the crucial and high-priority streets of of a city/town. This network lays the foundation for the integration of wireless Internet Protocol (IP) cameras at strategic locations, enabling real-time traffic monitoring and remote surveillance.

2 System Description and Hardware

The proposed system employs number of single-server–multiple clients-based street lighting system over the city/town under consideration. Figure 1 illustrates the envisioned layout of the IoT-based smart street lighting system designed for Cooch Behar Town, West Bengal. This layout focuses on a specific area within the town to showcase the proposed technology. For this purpose, the District Magistrate's office, located on the western bank of Sagar Dighi, is tentatively chosen to host the primary Internet Service Provider (ISP) and central internet sharing point. To facilitate expansive coverage, an outdoor router capable of dual-band (2.4 and 5.0 GHz) operation is employed. This router provides a WiFi radius extending up to 500 m (depicted by the red circle surrounding the main ISP in Fig. 1), achieved through the use of four dual-band WiFi broadcasting antennas. These antennas are optimally positioned at an elevation of 30–50 m above the ground, typically mounted on a tower installed atop the District Magistrate's office roof. Within the coverage area of the primary ISP router, all smart street lights establish an internet connection via this WiFi network. Furthermore, the ISP signal is extended to other repeater routers, such as the one positioned atop the high lighting mast at Station Chowpathi in Cooch Behar, through point-to-point (P2P) wireless links. Each repeater router employs a similar mechanism as the main ISP router, effectively propagating the WiFi network to the surrounding regions. This interconnected architecture forms the backbone of the smart street lighting system, fostering seamless connectivity and efficient coverage expansion throughout the designated area.

The sections between the main ISP router and the repeater router coverage areas, which often create WiFi blind spots (as observed along the majority of Sunity Road), are addressed through the utilization of outdoor WiFi extenders, as depicted in Fig. 1. This strategic approach ensures WiFi coverage for intermediate roads. As a result, key thoroughfares across Cooch Behar town such as Kesab Road, Sunity Road, and Naranarayan Road, among others, fall within the WiFi coverage umbrella. Smart street lights establish connectivity with their nearest router (main ISP, repeater, or extender) through corresponding Service Set Identifier (SSID) and password configurations. This configuration culminates in the creation of a wireless local area network (WLAN), utilizing the IEEE 802.11 protocol for two-way data exchange between the smart street light nodes and the cloud server database. These smart street lights, enveloped by WLAN coverage, are indicated on Fig. 1 by yellow star symbols enclosed in red borders. However, a significant portion of Cooch Behar town remains

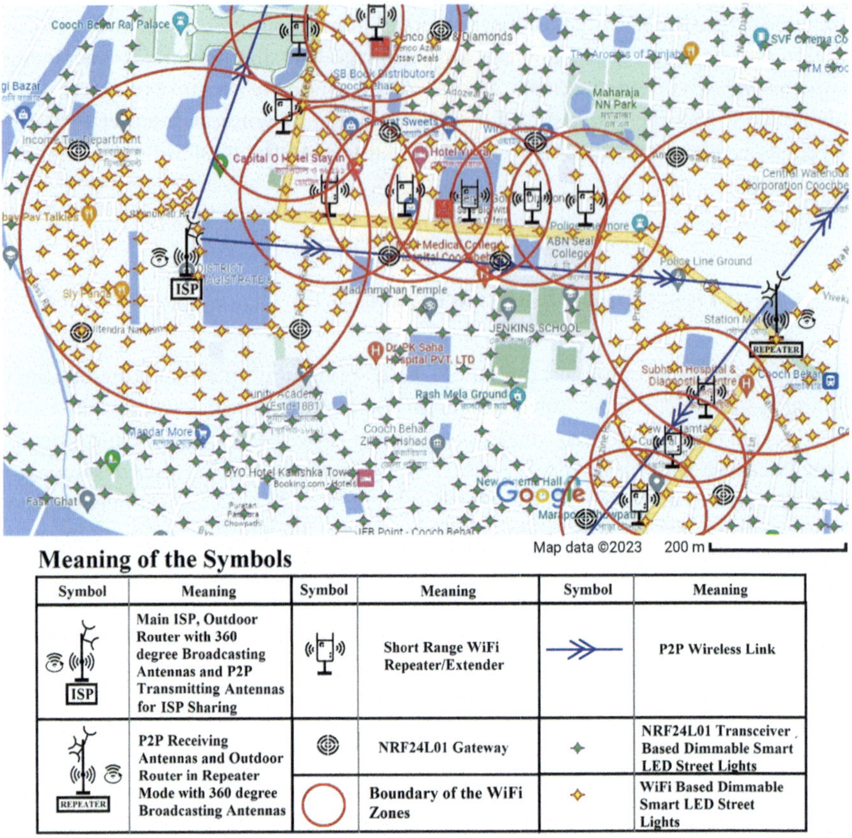

Fig. 1 Proposed layout of the IoT based smart street lighting system for Cooch Behar Town, West Bengal

beyond the reach of the WLAN. To establish bidirectional communication in these WiFi-deficient zones, a NRF24L01 gateway can be employed (located within a WiFi-enabled zone), along with NRF24L01 transceivers integrated into the smart street light nodes. Each NRF24L01 gateway has the capacity to connect with over three thousand (precisely 3125) NRF24L01 transceiver-equipped smart street light nodes. Consequently, the amalgamation of the proposed long-distance ISP sharing WLAN and the 2.4 GHz NRF24L01 mesh network ensures comprehensive access to the smart street lighting system for all street lights throughout the town.

A key advantage of the proposed technology lies in its seamless integration with existing street lights throughout the town, rendering the inclusion of these lights into the smart street lighting network straightforward–no replacement of the current LED lights is necessary. Each server node is equipped with a dedicated global positioning system (GPS) receiver to transmit precise location data (latitude, longitude, and altitude) of the respective street light to the cloud database during the initialization process. For automated operation, each node incorporates a light-dependent resistor (LDR) and a solid-state relay-based sensor. These sensors facilitate automatic switching on and off of the street light, respectively, based on ambient sunlight conditions during evening and dawn hours. An additional LDR sensor continuously monitors the street light's status, including on/off and dimming conditions, enabling remote malfunction detection. Enhancing functionality, an 180° fish-eye lens mounted no-infrared (NoIR) camera is used in the server node in order to achieve the instantaneous motion detection capability within the circular region-of-surveillance of the camera (around 250 m of radius) of each server node to detect nearby vehicles and pedestrians under 360° horizontal field-of-view of it. Figure 2 shows the block diagram of a server node. A server node is connected with multiple number of client-nodes through bidirectional WiFi socket communication channels. The block-diagram of a client-node is shown in Fig. 3. When all street lights are in power saving mode (at late-night), any motion detection in a server-camera instantaneously switches on the associated street light as well as all the street light client nodes corresponding to it through to socket channels. Thus, during late hours, any detected motion triggers an instant, full-brightness illumination response within the surrounding area. Conversely, when no activity is detected, the LED street lights are automatically dimmed to conserve power, achieved through a metal-oxide semiconductor field-effect transistor (MOSFET) based dimmer circuit. This intelligent approach optimizes power consumption substantially. Controlling these components, Raspberry Pi Zero 2W mini-computer and Pico 2W WiFi-enabled microcontroller are utilized within the smart street light server and client nodes respectively. Camera-based motion detection is not required in the client nodes, because those are strategically installed within the region-of-surveillance of the server camera. These mini-computer and microcontroller employ general-purpose input/output (GPIO) interfaces to connect with the various peripherals. An essential functionality includes a high-current solid-state relay controlled by GPIO, facilitating remote ON/OFF and dimming commands from the control room via a user console graphical user interface (GUI). This communication is facilitated through appropriate application programming interfaces (APIs), ensuring effective control and management of the smart street lights via the cloud database. A noteworthy attribute of the proposed smart nodes is their inherent programming that enables basic functions to operate autonomously even in cases where communication with the cloud server is lost. Consequently, any communication disruption with a specific street light cluster does not compromise the fundamental operations of the associated street lights within that cluster at any point in time.

The control room's smart street lighting user console GUI mirrors the configuration shown in Fig. 1. This graphical user interface facilitates seamless interaction with the real-time cloud-server database, enabling both data retrieval and updates. With a simple mouse cursor movement, users gain access to individual street light's on/off and dimming status. Manual intervention is also possible, allowing direct override of switch settings or dimming levels from the control room. To enhance operational insights, IoT-based weather stations can be strategically positioned across the town.

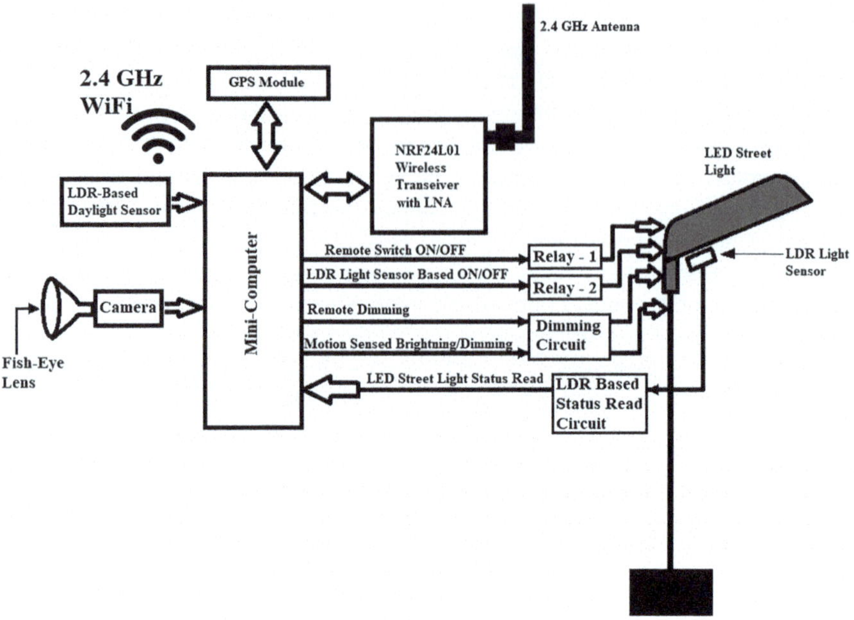

Fig. 2 Block-diagram of the hardware of a server-node

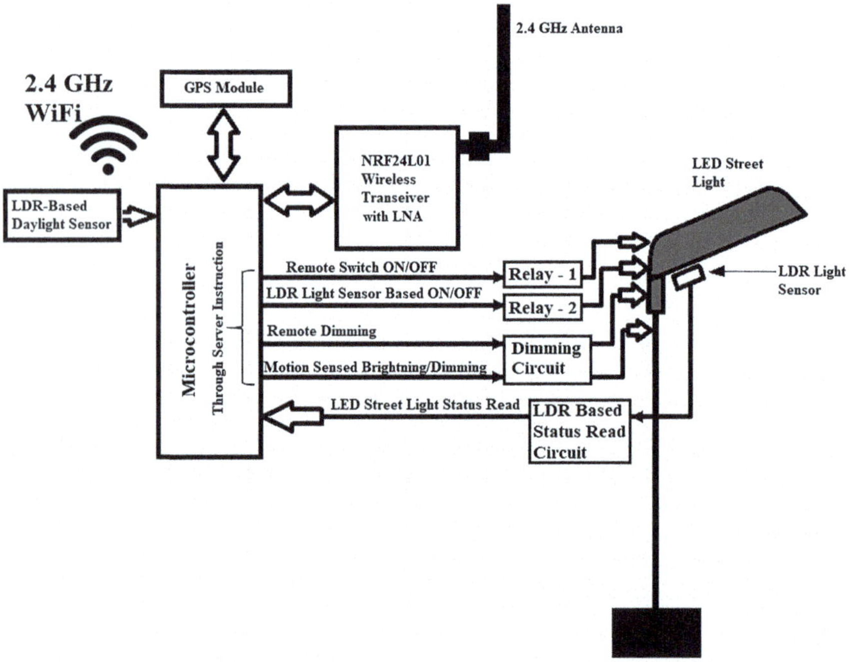

Fig. 3 Block-diagram of the hardware of a client-node

These stations are equipped with sensors for measuring attributes such as temperature, humidity, air pressure, moisture, and dust levels. By collecting real-time local weather and air quality data, these stations provide valuable information for decision-making. Moreover, the system's capabilities can be expanded by integrating AI-based wireless IP cameras within the WLAN's coverage area. These cameras can be selectively placed at key locations to facilitate traffic monitoring and remote surveillance, thereby enhancing overall safety and security measures.

3 Hardware Realization and Experimental Arrangements

In the preceding section, the author provided a comprehensive description of the hardware configurations for both the server and client nodes. At the IoT laboratory of the Department of Electronics and Communication Engineering, Cooch Behar Government Engineering College, Cooch Behar, the author successfully implemented an experimental setup, comprising a single server node and two client nodes. During the development and testing of both server and client nodes, the author utilized mock LED lights available within our laboratory. The server node is based on a Raspberry Pi Zero 2W, equipped with the Raspberry Pi Operating System (OS), and is registered on the Remote-it cloud service. This configuration allows for convenient remote access to the server from any personal computer using secure socket shell (SSH) or virtual network computing (VNC) services. A visual representation of the server prototype is presented in Fig. 4, with a closer view in Fig. 5. The client nodes, built around Raspberry Pi Pico 2W, are accessible from the remote end through "Thonny-Python" integrated development environment (IDE) and universal serial bus interface (USB). These client nodes are linked to another Raspberry Pi 4, Model B, featuring 4 GB of RAM, registered on the Remote-it cloud service. This arrangement is depicted in Fig. 6a, b. To enable remote viewing of the mock street light LEDs, the author incorporated live streaming capabilities through two cameras, one being a Raspberry Pi NoIR 8 MP version 2.0 camera, and the other being an 8 MP USB-interfaced web camera. Both cameras are connected to the Raspberry Pi 4-based remote server, to which the client nodes are also USB-interfaced. For a clearer perspective, snapshots of these live camera streams are provided in Fig. 7.

The author established the means for remote motion generation within the unoccupied laboratory by implementing a WiFi single-pole-dual-throw (SPDT) switch to control the mock LED lights. This arrangement is detailed in Fig. 8. When the WiFi SPDT switch is activated through a mobile application, transitioning from one state to another, it results in turning off the previously illuminated LED while turning on the previously unlit one, thus generating motion within the captured frame of the server camera. In order to provide a comprehensive understanding of the motion detection capabilities, the author offered a fish-eye view captured by the server camera in Fig. 9. The hardware setup and experimental arrangements described in this section form the foundation of our research, enabling the practical evaluation and validation of the proposed IoT-based smart lighting system.

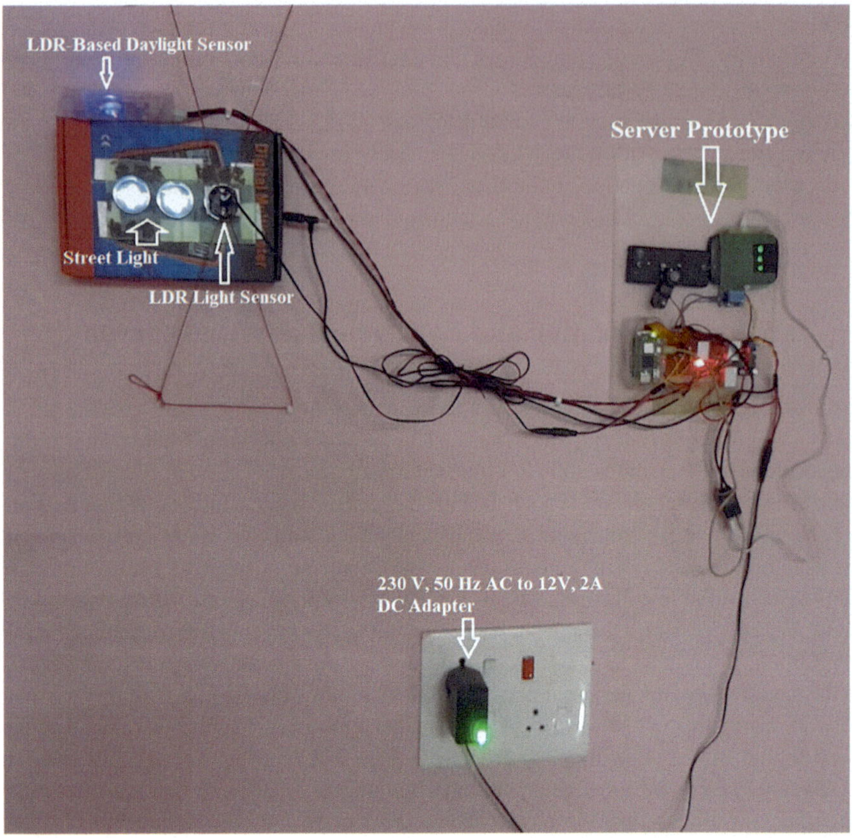

Fig. 4 Experimental setup for the server prototype

4 Software Development

The software implementation for the server node was carried out in the "Thonny-Python" Integrated Development Environment (IDE). Figure 10 illustrates the flowchart depicting the server algorithm. The server algorithm comprises four concurrent threads, each serving a distinct purpose: (i) LED Control Thread, (ii) Motion Capture Thread, (iii) Live Stream Thread (Optional), and (iv) Update Database Thread. The pivotal thread, the LED Control Thread, is scheduled for activation from 5:00 pm to 6:00 am, specifically during nighttime. This timing aligns with the presence of darkness, as the Street LED lights are automatically switched off during daylight hours by a light-dependent resistor (LDR)-controlled day-light sensing SPDT switch. Consequently, LED control is inactive from approximately 6:00 am to 5:00 pm throughout the year. However, it reactivates during daytime when darkness prevails due to overcast conditions or foggy atmospheres, triggered by the daylight sensor. On a standard day, the LED Control Thread maintains the

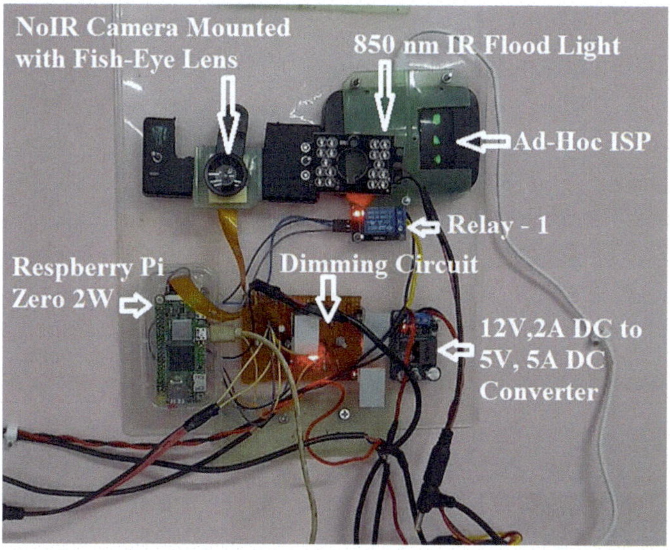

Fig. 5 Closer view of the server prototype

Street LED light in a fully illuminated state, operating in "Full-Brightness" mode between 5:00 pm and 6:00 pm. The LED Control Thread transitions into "Power-Saving" mode from 9:00 pm to 6:00 am. During this period, it initiates both the Motion Capture Thread and Live Stream Thread simultaneously. If no motion is detected within the server camera's field of view (utilizing an 850 nm infrared (IR) flood light for active night vision), the Street LED light remains off. When motion is detected in the server camera's captured frames, the Street LED light is promptly activated via the LED control relay (Relay 2). The Motion Capture Thread continually scans for motion within the camera's surveillance area. If motion persists, the Street LED light remains at full brightness. However, when motion is absent for at least one minute, a dimmer circuit reduces the street light's brightness to 30–50%. If no motion is detected for a period of 3–5 min, the Street LED light is entirely turned off by the LED Control Thread.

The Motion Capture Thread employs the "Successive Frame Subtraction" technique for motion detection within the server camera's surveillance area [4]. The Live Stream Thread is an optional feature that streams live footage of the server camera's surveillance area to the remote client. It is typically used for traffic surveillance purposes. The Update Database Thread remains active throughout the day, updating the status of the LED street light (ON/OFF/DIM) in the cloud database at 30-s intervals through an appropriate API. The current status of the LED street light is determined using an LDR-based LED-status reading circuit. The client nodes continuously monitor the status of their respective server LED lights and update the cloud database in sync with their own LED street lights through a bidirectional socket

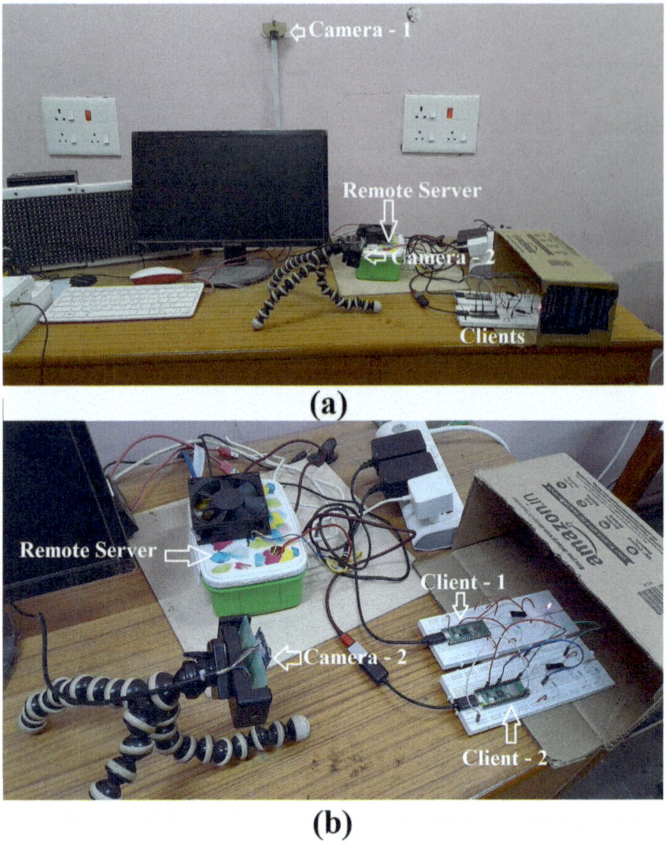

Fig. 6 Experimental setup for testing the clients

channel to the server. This process is straightforward and doesn't require a separate flowchart within this chapter.

5 Performance Evaluation

The manufacturer specified the field-of-view (FOV) for the fish-eye lens to be 180°. However, upon examination, it was determined that the practical FOV of the lens used was approximately 162°. The circular region-of-surveillance at the ground plane, denoted as R_{th}, for the server camera, when installed at a height 'h' from the ground plane as depicted in Fig. 11, can be theoretically calculated using the formula:

$$R_{th} = h \tan \theta_{th} \tag{1}$$

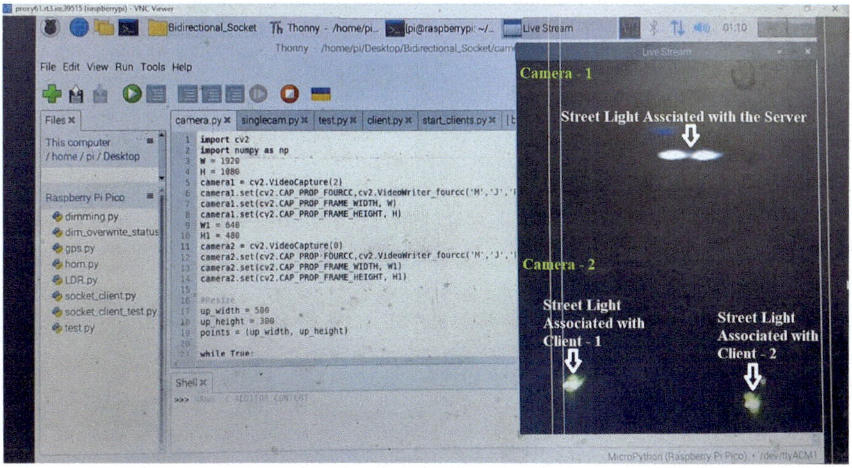

Fig. 7 Snapshots of the street lights associated with the server (captured from the Camera–1 of the testing remote server) and clients (captured from the Camera–2 of the testing remote server)

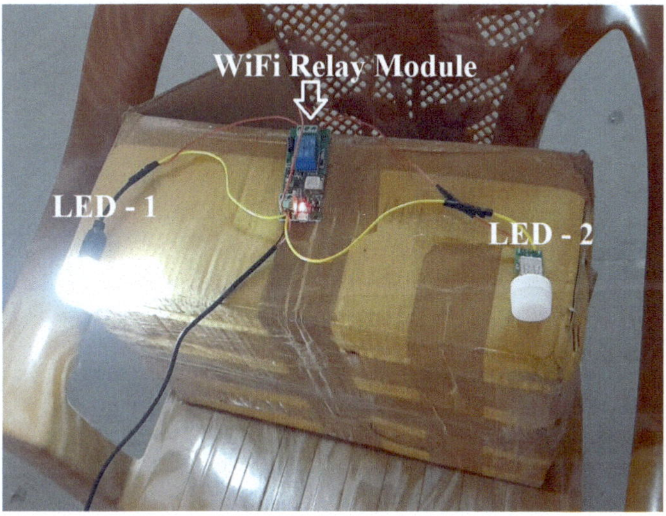

Fig. 8 Arrangement for motion generation from remote end

In practice, the motion capture range of the server camera is constrained by the video capturing resolution (640 × 480 pixels^2) and frame rate (30 frames per second (fps)). Thus, the practical motion detection range, denoted as R_{ex}, is limited to $R_{ex} \leq R_{th}$, and the corresponding angle is $2\theta_{ex} \leq 2\theta_{th}$, where $\theta_{th} = 81°$ (since FOV $= 2\theta_{th} = 162°$). The practical motion detection range is determined by the formula:

$$R_{ex} = h \tan \theta_{ex} \tag{2}$$

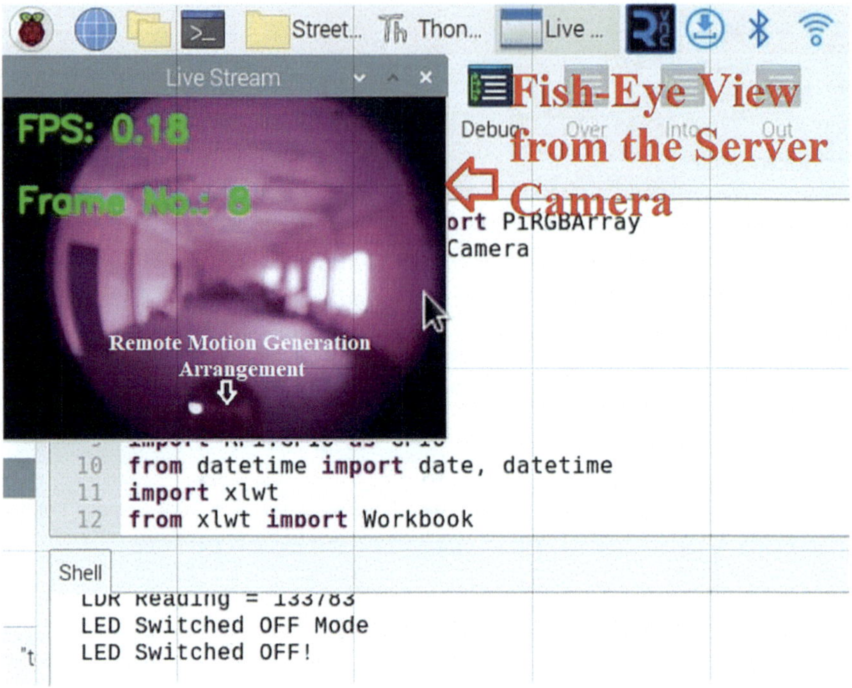

Fig. 9 Fish-eye view from the server camera for motion detection

The practical motion detection range of the server camera was measured for various camera heights. Figure 12 illustrates the variations in both theoretical and practical motion detection ranges with camera height. It was observed that up to 25 m of camera height, R_{th} and R_{ex} are nearly the same (approximately 157 m). However, as the camera height increases further, the practical motion detection range decreases compared to the theoretical value. Notably, the highest achievable R_{ex} value for a 640×480, 30 fps video stream with the NoIR v2.0 camera is approximately 250 m. Therefore, using a single server, a circular area with a maximum diameter of 500 m can be monitored for motion detection, and within this region-of-surveillance, any number of client nodes can be deployed.

The response time of the client nodes in relation to switching by the server was measured across different WLANs. The specifications of four distinct WLANs used for response time (t_r) measurements are detailed in Table 1. Figure 13 displays the measured client response times from 25 different trials for each of the considered WLANs. It is observed that the range of client response times falls within the range of 10 ms $\leq t_r \leq$ 152 ms, even under diverse network conditions. This response time range remains well below the threshold of human perceptibility, even under unfavorable network conditions.

The diagram in Fig. 14 illustrates the positioning of street lights for the final experiments. The server node is installed at a height of 50 m from the ground plane,

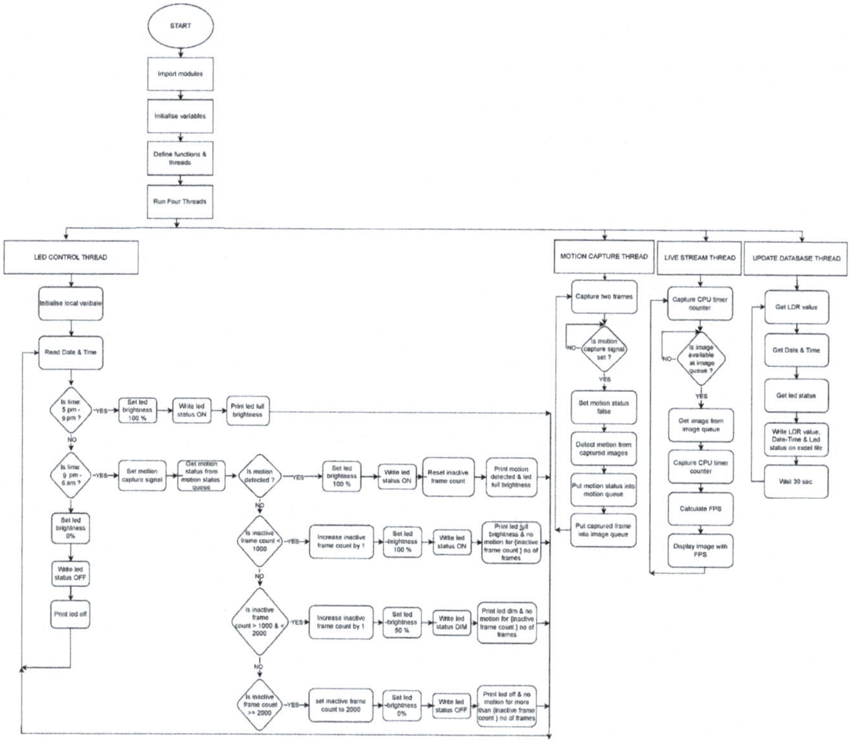

Fig. 10 Flowchart of the server-algorithm

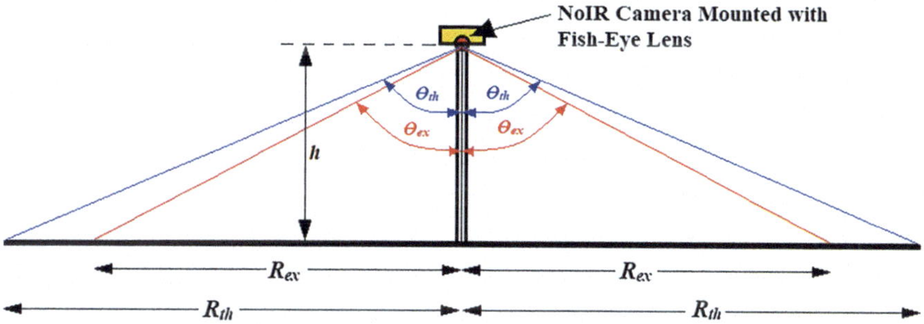

Fig. 11 Diagram illustrating the motion detection range of the server node

overseeing the control of eight 200 W street lights mounted on a high mast post. Each client node, belonging to client clusters 1–5, manages a 100 W street light mounted on 10-m-high poles. The bar graphs presented in Fig. 15 showcase the hour-wise percentage of energy savings in a day, measured on September 21, 2023, during the activation of the energy-saving mode from 9:00 pm to 06:00 am. Similar day-wise percentages of saved electrical energy (DWPoSE) were recorded over six

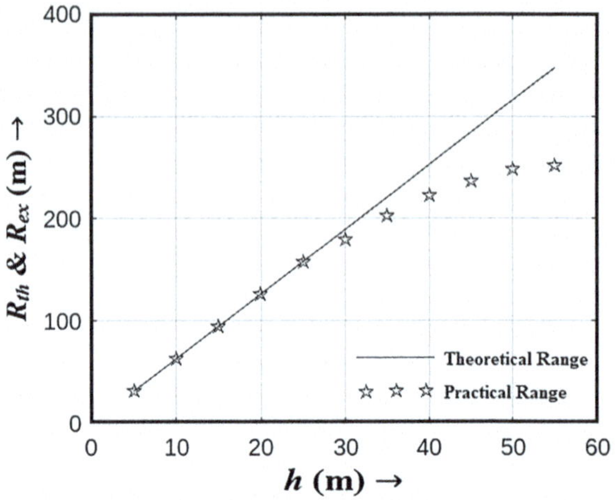

Fig. 12 Motion detection range versus height of the camera

Table 1 WLAN specifications

Network	IEEE standard	Frequency band (GHz)	Average upload speed (Mbps)	Average download speed (Mbps)	Average latency (ms)
WLAN1	802.11b	2.4	12.0	8.0	123.0
WLAN2	802.11b	2.4	40.3	34.6	90.0
WLAN3	802.11b	2.4	75.0	60.0	60.0
WLAN4	802.11b	2.4	122.3	101.7	15.0

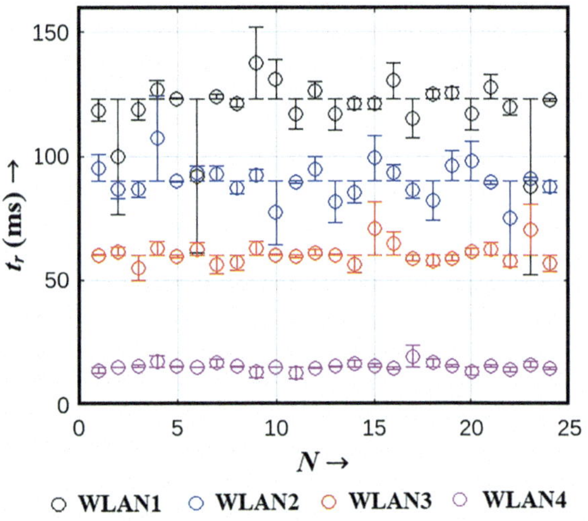

Fig. 13 Measured client response time versus trial number

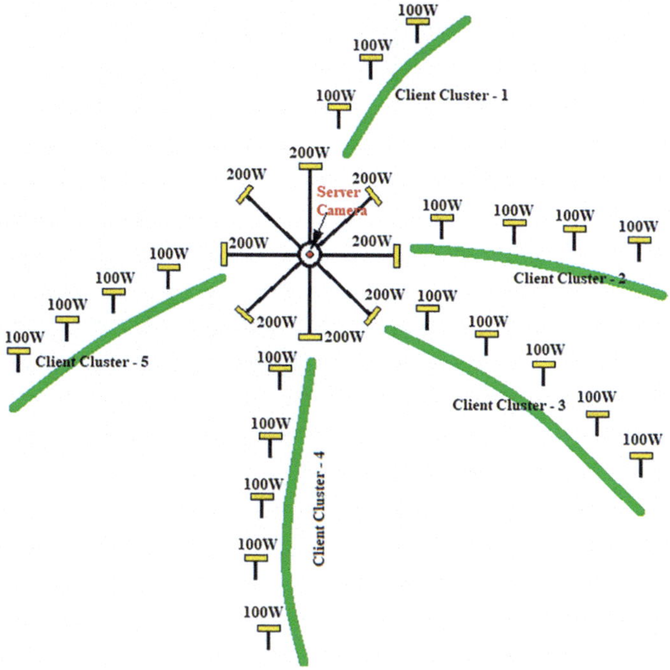

Fig. 14 Diagram showing the position of the street lights; server-node is installed at 50 m height from the ground plane to control total eight 200 W street lights and each client-node (within client-clusters 1–5) is controlling 100 W street light

months, from May 1, 2023, to October 31, 2023. From this data, the month-wise assigned electrical energy (AE), consumed electrical energy (CE), and percentage of saved electrical energy (MWPoSE) were calculated. These variations are depicted in Fig. 16. The assessment over a six-month period demonstrates that the proposed technology allows for approximately 53% of electrical energy savings with the 9:00 pm to 6:00 am active power-saving mode (50% dimming). Further enhancements can be achieved by adjusting the full brightness and dimming periods within the active power-saving mode and reducing the dimming percentage below 50%.

6 Advantages of the Proposed Technology

The proposed smart street lighting system offers several significant advantages:

(i) *Energy-Efficiency*: The system dramatically reduces power consumption by implementing motion camera-driven dimming and brightening functionality during late-night hours. This intelligent control ensures that energy is used

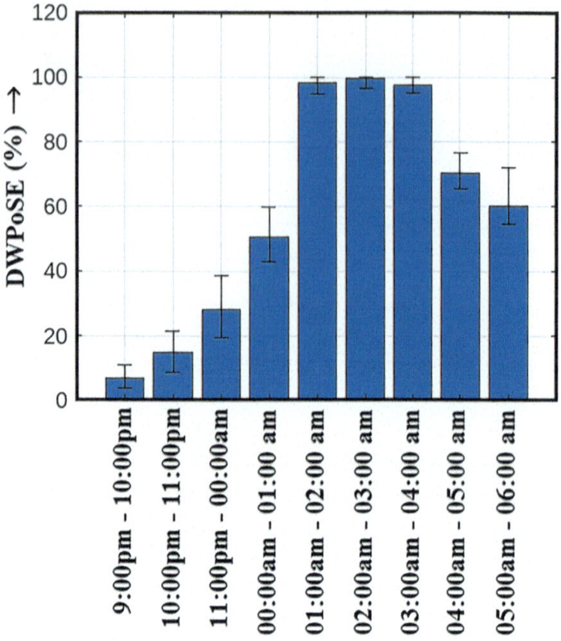

Fig. 15 Bar graphs showing the hour-wise percentage of energy saving in a day (21st September, 2023) while the energy-saving mode is activated during 9:00 pm–06:00 am

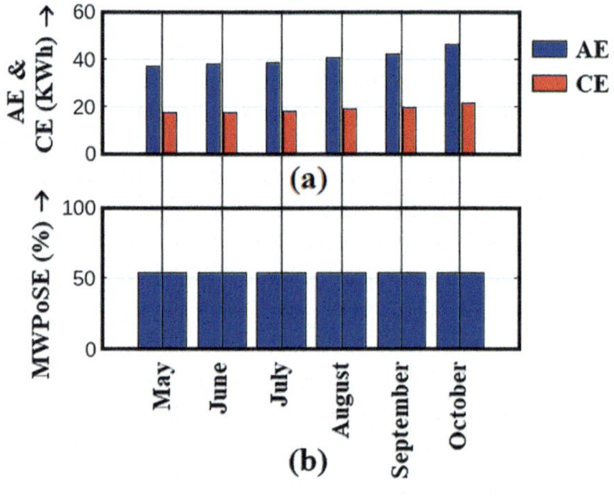

Fig. 16 Bar graphs showing the month-wise **a** assigned electrical energy (AE), consumed electrical energy (CE), and **b** percentage of saved electrical energy (MWPoSE) from 1st May, 2023 to 31st October, 2023

efficiently, leading to substantial cost savings and reduced environmental impact.

(ii) *Daytime Power Savings*: Wastage of power during daylight hours is minimized through the utilization of light sensors that trigger ON/OFF mechanisms, ensuring that street lights are active only when needed. This not only conserves energy but also reduces operational costs.

(iii) *Remote Access Control*: The system provides remote access control over the entire city's lighting network. This feature allows for easy monitoring, management, and adjustment of street light settings, contributing to enhanced operational efficiency.
(iv) *Real-Time Monitoring*: Real-time status updates enable the swift detection of malfunctioning street lights from a remote location. Maintenance and repairs can be promptly addressed, minimizing downtime and improving overall system reliability.
(v) *Seamless Integration*: The smart nodes seamlessly integrate with existing street lights, transforming them into intelligent units within the smart light network. This integration eliminates the need for costly new installations and accelerates the transition to energy-efficient lighting.
(vi) *Scalability and Versatility*: The system's architecture is designed to accommodate additional features, such as AI-powered IP cameras for traffic management and surveillance, as well as IoT-based weather stations. This flexibility allows for the expansion of the smart street light network to meet evolving urban needs while optimizing resource utilization.

7 Conclusion

In this research, the author introduced a novel approach for the enhancement of street lighting infrastructure in urban and suburban areas, as well as potentially along highways and railway tracks. This innovative approach integrates a long-distance Internet Service Provider (ISP) sharing strategy, offering an efficient solution for the deployment of a comprehensive smart lighting network. The system presents numerous advantages over existing technologies, especially those based on LoRaWAN, including lower installation and maintenance costs. The proposed smart street lights offer an array of essential functionalities, such as automated illumination, dynamic dimming, remote control, and real-time monitoring. By implementing ambient light sensors and motion detection capabilities, these lights intelligently respond to environmental conditions and movement, significantly reducing energy consumption. Moreover, the proposed system is highly cost-effective and seamlessly integrates with existing street lights, obviating the need for costly replacements.

The foundation of the proposed technology lies in the deployment of long-distance ISP sharing wireless local area networks (WLANs) across critical streets in cities and towns. This WLAN infrastructure supports the integration of wireless Internet Protocol (IP) cameras, enabling real-time traffic monitoring and remote surveillance, contributing to improved safety measures. Through a detailed hardware realization and experimental setup, the author successfully validated the proposed IoT-based smart lighting system. Practical measurements demonstrated that the proposed technology can efficiently cover a wide surveillance area, with motion detection capabilities even under low-light conditions. Real-world experiments confirmed that our system leads to substantial electrical energy savings, with up to 53% reduction in power consumption when the active power-saving mode is activated. The advantages of the proposed technology extend to its energy efficiency, day-night power savings, remote control capabilities, real-time monitoring, seamless integration, scalability, and versatility. Proposed system presents a significant contribution to the field of smart street lighting systems and large-scale electrical energy conservation, offering a practical and cost-effective solution for cities and towns aiming to optimize their lighting infrastructure.

This innovative approach marks a milestone in the path toward building smarter and more energy-efficient cities, improving the quality of life for urban and suburban dwellers while contributing to a greener and more sustainable future.

References

1. Wong W (2023) Smart streetlights help smart cities launch networking initiatives. Article Published in StateTech Magazine, 10th July, 2023. https://statetechmagazine.com/article/2023/07/smart-streetlights-help-smart-cities-launch-networking-initiatives. Accessed Oct 2023
2. Colclough G (2023) Smart lighting in cities. Article Published in EUSmartCities, pp 1–12. https://smart-cities-marketplace.ec.europa.eu/sites/default/files/2021-06/Smart%20Lighting%20Factsheet_0.pdf. Accessed Oct 2023
3. Colclough G (2019) Augmentation/up-gradation of existing LED streetlight to smart LED streetlights by adding the controller system including operation and maintenance in Ahmedabad city. Project Proposal of Smart City Ahmedabad Development Limited (SCADL), pp 1–90. https://csrbox.org/media/ahmd.pdf. Accessed Oct 2023
4. Ssandbac (2017) Raspberry Pi 3 Motion detection camera with live feed. Autodesk Instructables. https://www.instructables.com/Raspberry-Pi-Motion-Detection-Security-Camera/. Accessed Oct 2023

Chapter 5
Role of BERT Model for Sequential Text Classification in Biomedical Abstracts

Biplab K. Mandal, Purbayan Majumder, and Babul P. Tewari

Abstract In biomedical research, text classification plays a pivotal role when it comes to retrieving important data from large scientific abstract databases. Numerous healthcare fields stand to gain significant advantages from advancements in natural language processing (NLP), particularly within artificial neural networks. Sequential text categorization is highly influenced by Long Short-term memory networks (LSTMs) and recurrent neural networks (RNNs). However, issues such as exploding or vanishing gradients arise during training, impeding their ability to adequately capture long-range associations. Furthermore, these models might not be able to handle the complex semantic relationships included in text data. Recognizing these drawbacks highlights the need for creative solutions that push the limits of natural language processing (NLP) methods in order to more effectively handle the intricacies of biomedical text classification. Recognizing the limitations, efforts are focused on innovative ways to improve text classification in biomedical research, such as transformer topologies and attention processes. The purpose of this work is to investigate how well different LSTM architectures combined with BERT models might improve the understanding and categorization of biologically complicated sequential material in biomedical abstracts. The study emphasizes the dedication to improving techniques and utilizing the advantages of various models to get better results in the complex field of biomedical text classification. Our model employed RNN with LSTM architectures and two variations, comparing their performance with BERT, a state-of-the-art language model acclaimed for its exceptional contextual understanding. This approach explores the effectiveness of different architectures in sequential text classification for biomedical abstracts. The models underwent thorough training on three substantial datasets, yielding promising outcomes.

B. K. Mandal
Department of Information Technology, Asansol Engineering College, Kanyapur, Asansol, West Burdwan, West Bengal 713201, India

P. Majumder
Accenture Security, Sector 23, Gurugram, Haryana 122022, India

B. P. Tewari (✉)
Department of Computer Science and Engineering, Ghani Khan Choudhury Institute of Engineering and Technology, Malda, West Bengal 732141, India
e-mail: babul@gkciet.ac.in

A. Acharyya et al. (eds.), *Real-World Applications and Implementations of IoT*, Studies in Smart Technologies, https://doi.org/10.1007/978-981-97-8627-5_5

The comparative analysis revealed that our model excels, attaining state-of-the-art accuracy, precision, recall, and F1-score metrics in biomedical text classification. Notably, our model demonstrated robust generalization capabilities, indicating its potential for real-world applications. Utilizing advanced contextual comprehension and augmented hidden units, our adaptive BERT model outperformed competitors, highlighting its efficacy in extracting valuable insights from biomedical abstracts. This firmly establishes its status as a top-performing solution in biomedical research text classification. The models have been extensively trained on three large datasets to achieve promising results.

Keywords LSTM · RNN · BERT model · Text classification

1 Introduction

A branch of NLP known as sequential text classification [1] focuses on categorizing text [2] data that occurs in sequence, such as sentences within a document or statements in a conversation. Sequential text classification takes into account the contextual information supplied by the text units [3] that come before it, in contrast to classical text classification, which treats each text unit independently. By capturing the temporal interdependence and context inside the text sequence, this approach acknowledges that the texts contain valuable information that can improve the classification performance. The capacity of sequential text classification to use sequential information for increased accuracy is one of its main benefits. Sequential models are able to represent the underlying structure and dependencies found in the data by taking into account the placement and sequence of text units. This is especially helpful in applications like sentiment analysis [4, 5], where the interpretation of one section of the text might be influenced by the sentiment indicated in another. Since the order and structure of a document's content can influence its classification, sequential text classification is also useful for document classification.

Various methods for sequential text categorization have been explored, employing diverse models including Naive Bayes [6], Support Vector Machines [7], K-Nearest Neighbor [8], Logistic Regression [9], Recurrent Neural Networks (RNNs) [10] and Long Short-Term Memory (LSTM) networks [11]. RNNs and LSTMs are often preferred for sequential classification due to their ability to maintain internal memory states, enabling them to effectively capture context [12, 13]. This enables them to remember details from earlier text units and apply them to the categorization of later ones. Although LSTMs and RNNs have shown promising results in sequential text classification, there are certain limitations with these techniques. These models often struggle with capturing long-range dependencies and can suffer from vanishing or exploding gradients [14, 15] during training. To get more understanding of this matter, a thorough analysis has been conducted [16]. Additionally, they may lack the ability to handle the intricacies and semantic relationships present in textual data. BERT, a transformer-based model [17], overcomes these constraints by utilizing self-attention mechanisms and undergoing pre-training on extensive corpora, ultimately achieving state-of-the-art performance. Compared to RNNs and LSTMs, BERT offers better contextual understanding and handles long-range dependencies efficiently. This research focuses on comparing RNNs and LSTMs with BERT in classifying biomedical abstracts, vital for information retrieval and knowledge extraction in healthcare. Evaluating their effectiveness aims to provide insights into their performance and limitations.

2 Literature Survey

Text classification is a core task in NLP, used to automatically categorize or predict the class of unseen text documents, often utilizing supervised machine learning [1, 2, 6, 8]. Traditional methods of text categorization mainly focus on three aspects: feature engineering, feature selection, and the utilization of diverse machine learning algorithms to achieve classification goals [1]. Feature engineering involves transforming raw text data into numerical feature vectors that can be used as input for machine learning algorithms [18]. Hence, there is a need for increased effort to advance feature engineering within learning algorithms, aiming to enhance predictive outcomes and behaviors for better results. An extensive examination of methodologies, tools, and techniques utilized for feature engineering, aiming to enhance model (classifier) accuracy, has been discussed in [19]. The bag-of-words (BOW) feature representation [20] is indeed one of the most commonly utilized techniques in feature engineering for text classification tasks. The literature offers a number of feature selection overviews. As an instance, Chandrashekar and Sahin [21] classified feature selection methods into three groups: filter, wrapper, and embedding techniques. They provided a comprehensive overview of these methods. Focusing on multi-label classification, Pereira et al. [22] conducted an extensive survey and devised an innovative categorization of feature selection strategies. The objective of feature selection is to identify the most relevant features (words or phrases) that enhance the model's predictive capability while decreasing the dimensionality of the feature space [18]. Machine learning methods commonly utilize classifiers such as logistic regression (LR) [9], Naïve Bayes (NB) [6], and support vector machines (SVM) [7]. Logistic regression serves as a statistical technique for binary classification, while Naïve Bayes functions as a probabilistic classifier particularly suited for text classification and spam filtering. SVM, on the other hand, seeks to identify the optimal hyper plane to segregate data points across various classes within a high-dimensional space. However, these methods do not effectively address concerns like sparse information, scalability limitations, and cold-start hurdles. The use of matrix factorization (MF) techniques has been the subject of numerous studies to address some of these problems [23, 24]. Various neural models have been suggested for learning word representations. Neural networks show their excellent performance in various NLP tasks with the pre-trained [25] word embedding. The capacity of sequential text classification to use sequential information for increased accuracy in the area of sentiment analysis has been addressed in [4, 5]. The sentiment of a sentence is predicted by Socher et al. [26] using semi-supervised recursive auto-encoders. Text classification is essential for getting relevant information from massive scientific abstract databases in biological research. Recurrent neural networks (RNNs) [10] along with long short-term memory networks (LSTMs) [11] exhibit potential for sequential text classification; however, during training, issues such exploding or vanishing gradients arise, which impairs their ability to record long-range associations.

Moreover, it's possible that these models can't handle the intricate semantic links seen in text data. Acknowledging the drawbacks, efforts are concentrated on novel approaches, like transformer topologies and attention processes, to enhance text classification in biomedical research. The objective of this study is to explore potential enhancements in understanding and classification of physiologically complicated sequential content in biomedical abstracts that could result from combining various LSTM architectures with BERT models.

3 Existing Models

3.1 Recurrent Neural Network (RNN)

Neural networks capable of efficiently handling sequential data processing are referred to as recurrent neural networks (RNNs). RNNs have the capacity to capture temporal dependencies and contextual information embedded in sequential data, offering a distinct advantage over Feed Forward Neural Networks [3]. However, their performance is hampered by various challenges. One prominent issue is their struggle to capture long-term relationships due to the vanishing gradient problem [15]. This problem arises when gradients diminish exponentially over time, impeding the ability of RNNs to effectively learn and retain information for extended sequences. The inability to remember and use previous data over extended periods of time, which leads to information loss, is another issue. RNNs are characterized by their recurrent connections, enabling them to retain an internal memory or hidden state. During each time step, an RNN accepts input and adjusts its hidden state by incorporating both the current input and the prior hidden state [27]. The RNN architecture, depicted in Fig. 1, processes a given input time series $x = \{x_1, x_2, \ldots, x_t\}$ to compute the hidden state sequence $h = \{h_1, h_2, \ldots, h_t\}$ and the output sequence $y = \{y_1, y_2, \ldots, y_t\}$ iteratively. This capability allows the RNN to grasp the sequential structure of the data and generate predictions accordingly. RNNs are commonly employed as a foundational model in sequential text classification assignments because of their aptitude for capturing sequential dependencies.

There are several types of RNNs that have been proposed to address the limitations of standard RNNs. Some notable variants include Long-Short Term Memory (LSTM) [12] and the Gated Recurrent Unit (GRU) [13]. LSTM [12] incorporates memory cells and gating mechanisms to enhance the capture of long-term dependencies.

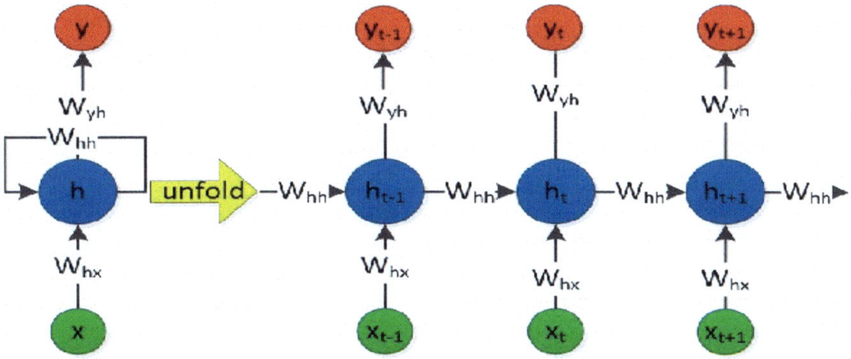

Fig. 1 The architecture of RNN

This also mitigates the vanishing gradient problem. Conversely, GRU simplifies the architecture by merging memory and hidden states, leading to a reduction in the number of parameters to be learned.

$$h_t = \sigma(W_{hx}x_t + W_{hh}h_{t-1} + b_h)$$

$$y_t = \sigma\left(W_{hy}h_t + b_y\right)$$

In this context, x_t denotes the input at time step t, h_t represents the hidden state at time step t, W_{hx} and W_{hh} are the weight matrices for the input-to-hidden and hidden-to-hidden connections, respectively. b_h denotes the bias term for the hidden state, W_{hy} is the weight matrix for the hidden-to-output connection, b_y is the bias term for the output, and σ signifies the activation, typically a sigmoid function.

3.2 Long Short-Term Memory (LSTM)

LSTM (Long Short-Term Memory) serves as a specialized form of RNNs engineered to combat the vanishing gradient issue and proficiently capture long-term dependencies within sequential data [12]. Its key feature is the memory cell, which selectively remembers or forgets information using input, forget, and output gates. This design empowers LSTM to preserve relevant information while filtering out extraneous details. The unit structure of LSTM, as depicted in Fig. 2 [28], encompasses four gates: the input modulation gate, input gate, forget gate, and output gate. LSTM finds utility across diverse domains, spanning natural language processing, speech recognition, machine translation, and sentiment analysis. In tasks involving sequential text classification, LSTM demonstrates proficiency in capturing extensive dependencies and contextual nuances within textual data. This renders it well-suited for endeavors such as sentiment analysis, text categorization, and language modeling.

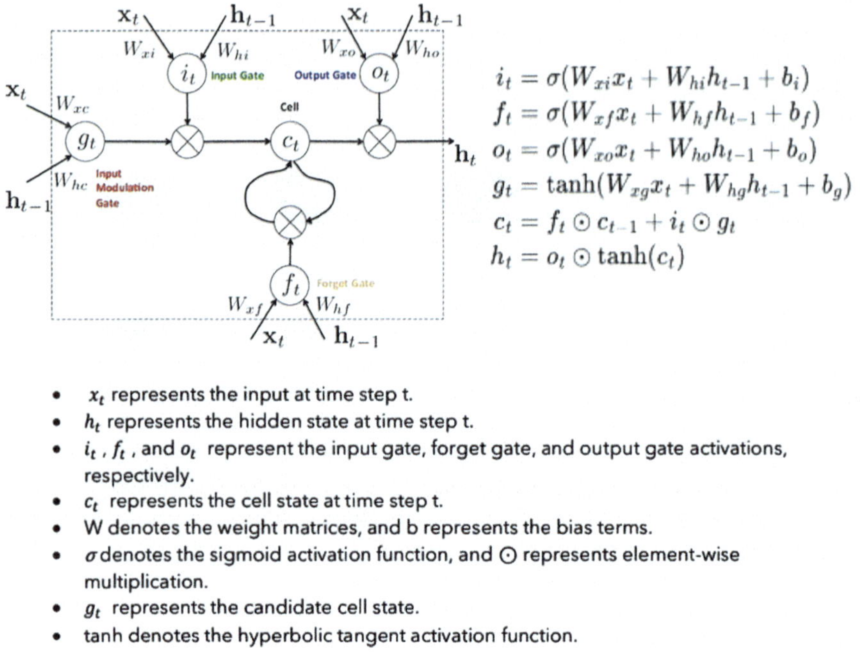

Fig. 2 Architecture of LSTM

As an experimental part of this work, two LSTM models were evaluated: one with GloVe embeddings and another with custom token embeddings and positional encodings [29]. Custom token embeddings, trained specifically on the dataset, capture task-specific semantics, while positional encodings encode word positions in the input sequence [12]. To incorporate positional information, custom positional encodings were added to input embeddings. These encodings, computed using sinusoidal functions, capture positional information continuously. Combining custom token embeddings and positional encodings enables the LSTM model to leverage task-specific semantics and positional information, improving performance in sequential text classification tasks.

$$E_{\text{custom}} = f_x(x_1, x_2, \ldots, x_3)$$

$$E_{\text{pos}}[p,\ 2_i] = \sin\left(\frac{p}{10000^{2i/d}}\right)$$

$$E_{\text{pos}}[p,\ 2_i + 1] = \cos\left(\frac{p}{10000^{2i/d}}\right)$$

Custom Embeddings Positional Embeddings

3.3 Bidirectional Encoder Representations from Transformers (BERT)

A highly effective open-source machine learning framework for natural language processing (NLP) known as BERT (Bidirectional Encoder Representations from Transformers) [17] was developed by Google. Its cutting-edge performance on numerous NLP tasks has attracted a lot of interest and made it quite popular. Unlike traditional models like LSTM and RNN, BERT introduces a transformer-based architecture that captures contextualized word representations in both directions. Having been pre-trained on an extensive volume of unlabeled text sourced from the internet, BERT has acquired profound and comprehensive language representations. BERT's versatility shines through its capability to handle various natural language processing tasks such as text classification [30], named entity recognition [31], question answering, and machine translation. This adaptability stems from its utilization of unsupervised pre-training, complemented by task-specific fine-tuning. This approach enables BERT to excel in various domains and adapt to different downstream applications.

In fact, BERT diverges from the conventional sequence-to-sequence model with encoder-decoder architecture. Instead, it employs the transformer architecture [32], which is entirely encoder-based. Transformer architecture like BERT, has layers for encoders and decoders, however BERT only uses the encoder portion shown in Fig. 3. In jobs like machine translation, where the model must produce a series of output tokens depending on an input sequence, the decoder is usually utilized. The internal architecture of BERT consists of transformer blocks that incorporate self-attention mechanisms. BERT leverages attention mechanisms, enabling it to selectively attend to diverse words in the input sequence and effectively capture their contextual relationships. The attention scores are computed through the self-attention mechanism, and the resulting contextualized representations are fed into multiple layers of transformers.

The basic equations for understanding BERT's mechanisms such as self-attention, feed-forward neural network layers, layer normalization, and residual connections are given below.

$$\text{Attention}(Q, K, V) = \text{softmax}\left(\frac{QK^T}{\sqrt{d_k}}\right)v$$

- Q, K, V: Query, Key, and Value matrices.
- dk: Dimensionality of the Key matrix.

$$\text{FFN}(x) = \max(0, xW_1 + b_1)W_2 + b_2$$

- x: Input vector.
- W_1, b_1: Weights and biases of the first linear transformation.

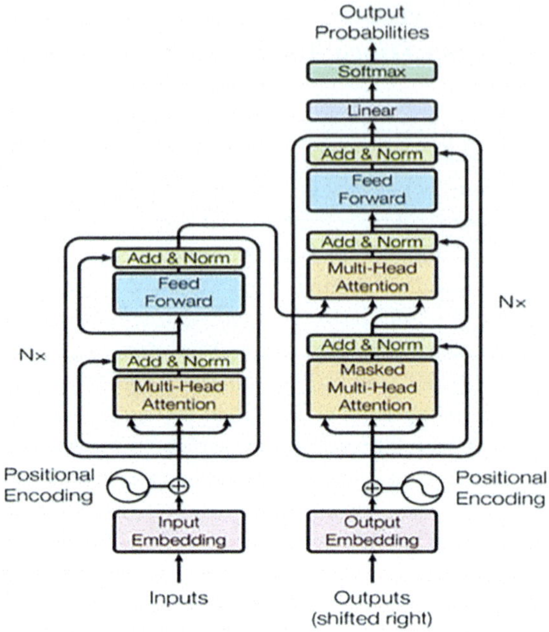

Fig. 3 Transformer—model architecture for BERT

- W_2, b_2: Weights and biases of the second linear transformation.

$$\text{LayerNorm}(x) = \gamma\left(\frac{x - u}{\sqrt{\sigma^2 + \varepsilon}}\right) + \beta$$

- x: Input vector.
- γ: Scaling parameter.
- μ: Mean of the input vector.
- σ^2: Variance of the input vector.
- ε: Small constant for numerical stability.
- β: Shifting parameter.

$$\text{Residual}(x) = \text{LayerNorm}(x + \text{Sublayer}(x))$$

- x: Input vector.
- Sublayer (x): Sublayer operation applied to the input vector.

4 Role of BERT Models and Proposed Techniques

BERT is a versatile and task-specific pre-training model that employs masked language modeling (MLM) [25] during pre-training. BERT overlooks the interdependence among predicted tokens, utilizing metrics tailored for an immediately applied database. The model is utilized for document and patent classification because of its perfect accuracy. By optimizing techniques, it is accomplished more than 100 pre-trained languages in the BERT model are helpful for various projects. In our context, BERT presents advantages compared to LSTM and RNN models in the analysis of biomedical text. Its capacity to grasp contextual information and manage long-range dependencies renders it particularly suitable for comprehending and categorizing biomedical text. Pre-training on a large corpus of biomedical literature enhances its domain-specific knowledge [17]. BERT's internal architecture comprises transformer blocks with self-attention mechanisms [32], enabling effective capture of contextual relationships among words in input sequences. Attention scores are computed through self-attention mechanisms, and resulting contextualized representations are processed through multiple transformer layers [17]. We investigate the efficacy of using BERT models in conjunction with LSTM model variants for sequential text categorization of biomedical abstracts. We used two different variants of RNN with LSTM-based architectures in our proposed models. Subsequently, we evaluated the performance of these models in comparison to BERT, an advanced language model celebrated for its exceptional contextual comprehension.

5 Experiment

5.1 *Word Embedding*

Word embeddings, also known as word representations [29], are numerical vector representations of words that capture semantic and syntactic information about their meanings. These embeddings find extensive application in various natural language processing tasks, such as text classification, machine translation, and information retrieval. They furnish a concise and meaningful portrayal of words, empowering algorithms to discern relationships and similarities among words within a high-dimensional space. The embedding used specifically here is GloVe (Global Vectors for Word Representation) [29]. GloVe stands as a widely recognized word embedding method, that leverages global word co-occurrence statistics to learn word representations. The glove.6b.50d embedding consists of word vectors with a dimensionality of 50, trained on a large corpus which means 6 billion tokens and 400,000 vocab with 50d vectors. These embeddings encapsulate both contextual information and semantic relationships among words, thereby empowering our model to comprehend word meanings within their surrounding context [33]. Compared to other word embeddings such as Word2Vec [34] and fastText [35], GloVe has shown significant improvements in capturing semantic relationships, preserving semantic relationships, and preserving syntactic regularities. GloVe embedding demonstrates superior performance across a range of natural language processing tasks and has gained widespread adoption in both research and industry settings. Additional prominent word embedding methods include Word2Vec, leveraging neural networks to acquire word representations, and fastText, which integrates sub-word information into word embedding. However, our choice of GloVe embedding aligns well with the requirements of our model and has demonstrated its effectiveness in previous studies.

5.2 *Optimizers*

In our research, we compared three optimization functions: Adagrad, Adam, and RMSprop [36]. Adagrad and RMSprop didn't dynamically decay the learning rate over time, causing them to struggle in adapting to the varying importance of words or phrases in sequential text classification. This led to suboptimal performance. Adam addressed this issue by using adaptive learning rates based on estimated gradient moments. However, when applied to our RNNs, Adam faced challenges with squared gradients dominating the learning process, leading to learning rate decay and convergence issues.

$$\text{Learning rate} = \frac{\text{Initial Learning Rate}}{\sqrt{\text{Accumulated Squared Gradients}} + \text{Epsilon}}$$
$$\text{Output} = \text{Input} * \text{Mask}$$

Learning rate of Adagrad Algorithm

$$m_t = \beta_1 * m_{t-1} + (1 - \beta_1) * g_t$$
$$v_t = \beta_2 * v_{t-1} + (1 - \beta_2) * g_t^2$$
$$\hat{m}_t = \frac{m_t}{1 - \beta_1^t}$$
$$\hat{v}_t = \frac{v_t}{1 - \beta_2^t}$$
$$\theta_{t+1} = \theta_t - \frac{\text{Learning Rate}}{\sqrt{\hat{v}} + \varepsilon} + \hat{m}_t$$

Adam Algorithm

Adam combines the benefits of Adagrad and RMSprop while addressing their limitations. It computes first and second moment estimates of gradients (m_t and v_t), controlled by hyper parameters β_1 and β_2 for decay rates. A small constant ε prevents division by zero. The adaptive learning rate is based on second moment estimates and scaled by the first moment estimates.

5.3 *Dataset*

Abstracts from biomedical literature, obtained from sources like PubMed, PubMed Central (PMC), and ClinicalTrials.gov, provided valuable to us for training and evaluating models for biomedical text classification. These abstracts followed a structured format, including sections like objective, background, methods, results, and conclusion. By extracting and labeling these sections shown in Fig. 4, we trained our LSTM, RNN, and BERT models to understand and classify distinct aspects of our research studies. This helped us to capture vital information and enhanced the models' performance in tasks like information retrieval and summarization in the biomedical domain. These datasets were split into the sizes of 20,000 and 200,000 abstracts and then were labeled which helped us to train our model effectively.

5.4 *Preprocessing*

5.4.1 Tokenization and Stop Words Removal

In sequential text classification of biomedical abstracts, preprocessing techniques like tokenization, labeling, and stop words removal were applied [5, 23]. Tokenization breaks text into individual tokens for analysis, labeling assigns categories to

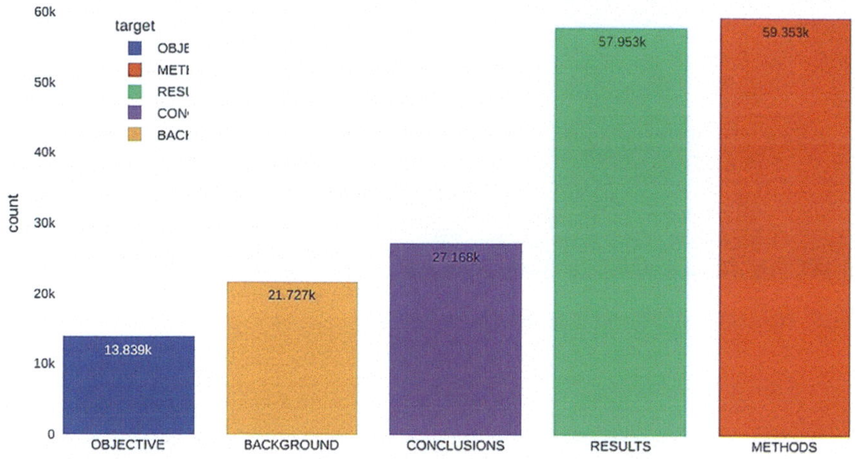

Fig. 4 Details of the processed dataset

tokens, aiding in pattern recognition [5]. Stop words removal eliminates common words to reduce noise and enhance efficiency [23]. These techniques improved accuracy, feature representation, and overall performance, leading to more precise and informative results.

5.4.2 One-Hot Encoding and Label Encoding

During preprocessing for sequential text classification of biomedical abstracts, two encoding techniques, one-hot and label encoding, were used [7, 37]. One-hot encoding transformed categorical variables into binary vectors, providing a unique binary representation for each category [7]. Similarly, label encoding assigned distinct numerical labels to categories, transforming them into numerical values [37]. These methods ensure compatibility between data and classification models, facilitating effective learning and accurate predictions. Leveraging one-hot and label encoding enhanced data representation, streamlined model training, and improved classification performance (Fig. 5).

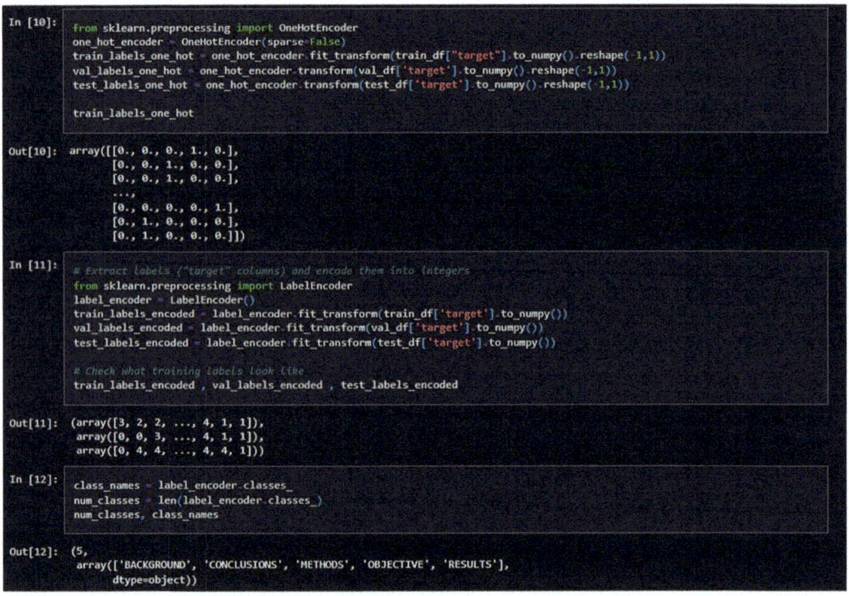

```
In [10]:
from sklearn.preprocessing import OneHotEncoder
one_hot_encoder = OneHotEncoder(sparse=False)
train_labels_one_hot = one_hot_encoder.fit_transform(train_df["target"].to_numpy().reshape(-1,1))
val_labels_one_hot = one_hot_encoder.transform(val_df['target'].to_numpy().reshape(-1,1))
test_labels_one_hot = one_hot_encoder.transform(test_df['target'].to_numpy().reshape(-1,1))

train_labels_one_hot

Out[10]: array([[0., 0., 0., 1., 0.],
       [0., 0., 1., 0., 0.],
       [0., 0., 1., 0., 0.],
       ...,
       [0., 0., 0., 0., 1.],
       [0., 1., 0., 0., 0.],
       [0., 1., 0., 0., 0.]])

In [11]:
# Extract labels ("target" columns) and encode them into integers
from sklearn.preprocessing import LabelEncoder
label_encoder = LabelEncoder()
train_labels_encoded = label_encoder.fit_transform(train_df['target'].to_numpy())
val_labels_encoded = label_encoder.fit_transform(val_df['target'].to_numpy())
test_labels_encoded = label_encoder.fit_transform(test_df['target'].to_numpy())

# Check what training labels look like
train_labels_encoded , val_labels_encoded , test_labels_encoded

Out[11]: (array([3, 2, 2, ..., 4, 1, 1]),
 array([0, 0, 3, ..., 4, 1, 1]),
 array([0, 4, 4, ..., 4, 4, 1]))

In [12]:
class_names = label_encoder.classes_
num_classes = len(label_encoder.classes_)
num_classes, class_names

Out[12]: (5,
 array(['BACKGROUND', 'CONCLUSIONS', 'METHODS', 'OBJECTIVE', 'RESULTS'],
       dtype=object))
```

Fig. 5 Process of one-hot encoding and label encoding

5.5 Training Process

The datasets underwent automatic partitioning into three segments: 70% for training, 20% for validation, and 10% for testing. This splitting process was performed to ensure that the data was divided appropriately for training, evaluation, and testing purposes (Table 1). By adhering to this standard procedure, it enabled a systematic and trustworthy evaluation of the model's performance on unseen data.

5.6 Hyperparameters

In this study, the hyperparameters were set to a learning rate of 1×10^{-4} and 1×10^{-5} activation function as ReLU, optimizer as Adam, and softmax as the final layer activation function shown in Table 2.

Table 1 Dataset splits

Dataset sizes	Training	Validation	Testing
20,000	14,000	4,000	2,000
200,000	140,000	40,000	20,000

Table 2 Hyperparameters of the model

Model	Optimizer	Loss function	Hidden activation function	Output activation function	Learning rate
LSTM model (custom token embedding)	Adam	Cross entropy loss	ReLU	Softmax	1×10^{-4}
LSTM model (glove embedding)	Adam	Cross entropy loss	ReLU	Softmax	1×10^{-4}
LSTM model (positional encoding)	Adam	Cross entropy loss	ReLU	Softmax	1×10^{-5}
Proposed Bert model	Adam	Cross entropy loss	ReLU	Softmax	1×10^{-5}

6 Result Analysis

The experimental results demonstrate the significant impact of embeddings and encodings on the performance metrics of the LSTM models. Notably, Model 1 and Model 2 exhibited remarkable performance during the training phase, leveraging custom token embeddings and GloVe embeddings, respectively. Model 3, with its increased number of layers and the incorporation of positional encoding, demonstrated stability across both embeddings shown in Table 3. However, the standout performer was BERT, leveraging its robust self-attention mechanism and deeper architecture. With its superior contextual understanding and increased hidden units, BERT outperformed all other models. Additionally, the inclusion of a dropout layer effectively mitigated over fitting across all models. These findings highlight the pivotal role of strong embeddings, encodings, and architectural design in improving model performance for sequential text classification tasks, which is graphically shown in Fig. 6.

Table 3 Result analysis

Model	Training	Validation
LSTM model (custom token embedding)	81.7	73.2
LSTM model (glove embedding)	87.1	72.4
LSTM model (positional encoding)	80.1	77.79
Proposed Bert model	84.0	88.0

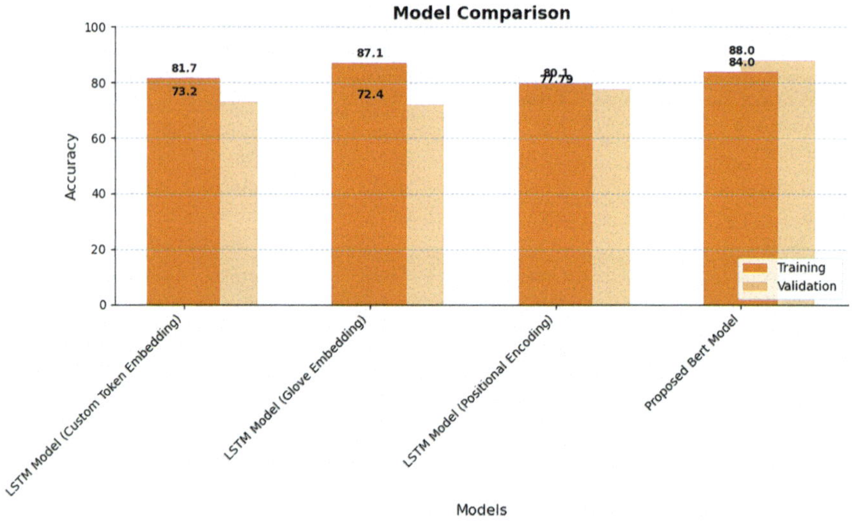

Fig. 6 Result analysis of models

7 Conclusion

This study presents a model that compares the performance of two versions of an RNN with LSTM architectures to that of BERT, a cutting-edge language model widely recognized for its remarkable contextual comprehension. Three large datasets were used to thoroughly train the model, with encouraging results. Our model outperformed the competition, as demonstrated by comparison analysis, which produced state-of-the-art metrics for biomedical text categorization, including accuracy, precision, recall, and F1-score. Furthermore, our model demonstrates strong generalization abilities, indicating its practical utility in biomedical text classification.

References

1. Lai S, Xu L, Liu K, Zhao J (2015) Recurrent convolutional neural networks for text classification. In: AAAI conference on artificial intelligence, vol 29, no 1
2. Chen G, Ye D, Xing Z, Chen J, Cambria E (2017) Ensemble application of convolutional and recurrent neural networks for multi-label text categorization. In: International joint conference on neural networks (IJCNN), pp 2377–2383
3. Cho K, Merriënboer BV, Gulcehre C, Bahdanau D, Bougares F, Schwenk H, Bengio Y (2014) Learning phrase representations using RNN encoder-decoder for statistical machine translation. In: Conference on empirical methods in natural language processing (EMNLP), pp 1724–1734
4. Socher R, Perelygin A, Wu J, Chuang J, Manning CD, Ng AY, Potts C (2013) Recursive deep models for semantic compositionality over a sentiment treebank. In: Proceedings of the 2013 conference on empirical methods in natural language processing, pp 1631–1642
5. Mass AL, Daly RE, Pham PT, Huang D, Ng AY, Potts C (2011) Learning word vectors for sentiment analysis. In: Proceedings of the 49th annual meeting of the association for computational linguistics: human language technologies
6. Jiang L, Li C, Wang S, Zhang L (2016) Deep feature weighting for naive Bayes and its application to text classification. Eng Appl Artif Intell 52:26–39
7. Fanjin M, Ling H, Jing T, Xinzheng W (2017) The research of semantic kernel in SVM for Chinese text classification. In: Proceedings of the 2nd international conference on intelligent information processing. ACM, p 8
8. Azam M, Ahmed T, Sabah F, Hussain MI (2018) Feature extraction based text classification using K-nearest neighbor algorithm. IJCSNS Int J Comput Sci Netw Secur 18(12):95–101
9. Pranckevicius T, Marcinkevičius V (2016) Application of logistic regression with part-of-the-speech tagging for multi-class text classification. In: 2016 IEEE 4th workshop on advances in information, electronic and electrical engineering (AIEEE)
10. Gao M, Li T, Huang P (2018) Text classification research based on improved Word2Vec and CNN. In: International conference on service-oriented computing. Springer, Cham, pp 126–135
11. Johnson R, Zhang T, Doshi-Velez F (2016) Supervised and semi-supervised text categorization using LSTM for region embeddings. In: Proceedings of the 2016 conference on empirical methods in natural language processing (EMNLP)
12. Hochreiter S, Schmidhuber J (1997) Long short-term memory. Neural Comput 1735–1780
13. Chung J, Gulcehre C, Cho K, Bengio Y (2014) Empirical evaluation of gated recurrent neural networks on sequence modeling. In: NIPS 2014 workshop on deep learning
14. Yan Y, Wang Y, Gao WC, Zhang BW, Yang C, Yin XC (2018) LSTM2: multi-label ranking for document classification. Neural Process Lett 47(1):117–138

15. Bengio Y, Simard P, Frasconi P (1994) Learning long-term dependencies with gradient descent is difficult. IEEE Trans Neural Netw 5(2):157–166
16. Pascanu R, Mikolov T, Bengio Y (2013) On the difficulty of training recurrent neural networks. In: Proceedings of the 30th international conference on machine learning. J Mach Learn Res 28
17. Devlin J, Chang MW, Lee K, Toutanov K (2019) BERT: pre-training of deep bidirectional transformers for language understanding. In: Nations of the Americas chapter of the association for computational linguistics, pp 4171–4186
18. Verdonck T, Baesens B, Oskarsdottir M, Broucke SV (2021) Special issues on feature engineering editorial. J Mach Learn
19. Rawat T, Khemchandani V (2017) Feature engineering (FE) tools and techniques for better classification performance. Int J Innov Eng Technol (IJIET) 8(2)
20. Qader WA, Ameen MM, Ahmed BI (2019) An overview of bag of words;importance, implementation, applications, and challenges. In: International engineering conference
21. Chandrashekar G, Sahin F (2016) A survey on feature selection methods. Comput Electr Eng 40(1):16–28
22. Pereira RB, Plastino A, Zadrozny B, Merschmann LH (2016) Categorizing feature selection methods for multi-label classification. Artif Intell Rev 49(1):57–78
23. Sinha BB, Dhanalakshmi R (2022) DNN-MF: deep neural network matrix factorization approach for filtering information in multi-criteria recommender systems. Neural Comput Appl 34(23)
24. Fu M, Qu H, Yi Z, Lu L, Liu Y (2018) A novel deep learning-based collaborative filtering model for recommendation system. IEEE Trans Cybern
25. Dong L, Yang N, Wang W, Wei F, Liu X, Wang Y, Gao J, Zhou M, Hon HW (2019) Unified language model pre-training for natural language understanding and generation. In: Advances in neural information processing systems (NeurIPS), vol 32
26. Socher R, Pennington J, Huang EH, Ng AY, Manning CD (2011) Semi-supervised recursive autoencoders for predicting sentiment distributions. In: Empirical methods in natural language processing, pp 151–161
27. Zheng J, Xu C, Zhang Z, Li X (2107) Electric load forecasting in smart grid using long-short-term-memory based recurrent neural network. In: 51st annual conference on information sciences and systems (CISS), pp 1–6
28. Chen G (2016) A gentle tutorial of recurrent neural network with error backpropagation
29. Pennington J, Socher R, Manning CD (2014) GloVe: global vectors for word representation. In: Proceedings of the 2014 conference on empirical methods in natural language processing (EMNLP)
30. Zhang X et al (2015) Character-level convolutional networks for text classification. In: Advances in neural information processing systems (NIPS), pp 649–657
31. Duppati SK, Babu A (2023) Named entity recognition for english language using deep learning based bi directional LSTM-RNN. Int J Recent Innov Trends Comput Commun 11(5):330–337
32. Vaswani A, Shazeer N, Parmar N, Uszkoreit J, Jones L, Gomez AN, Kaiser L, Polosukhin I (2017) Attention is all you need. In: 31st conference on neural information processing systems (NIPS)
33. Mikolov T, Sutskever I, Chen K, Corrado GS, Dean J (2013) Distributed representations of words and phrases and their compositionality. In: Advances in neural information processing systems (NIPS)
34. Mikolov T, Chen K, Corrado G, Dean J (2013) Efficient estimation of word representations in vector space
35. Bojanowski P, Grave E, Joulin A, Mikolov T (2017) Enriching word vectors with subword information. Trans Assoc Comput Linguist 5:135–146
36. Kingma DP, Ba JL (2015) Adam: a method for stochastic optimization. In: International conference on learning representations
37. Collobert R, Weston J, Bottou L, Karlen M, Kavukcuoglo K (2011) Natural language processing (almost) from scratch. J Mach Learn Res 12:2493–2537

Chapter 6
Wi-Fi Enabled IoT Communications: A Low Power Consumption Perspective

Babul P. Tewari and Poulomi Mukherjee

Abstract Internet of Things (IoT) based communications are expected to be energy efficient. This is because IoT devices are mostly battery operated and frequent replacement of batteries is not desirable. From this perspective, energy efficient IoT communications have drawn much attention, which is a key area of future IoT communications. Wi-Fi technology has continuously evolved from its legacy standards through high speed communications of IEEE 802.11ac also known as Wi-Fi5 and nowadays IEEE 802.11ax known as Wi-Fi6. Such Wi-Fi technologies have also proved their suitability for heterogeneous communications, an important requirement of future wireless communications. In this paper, we have addressed the key roles to be played by IEEE 802.11 WLAN technology in implementing IoT-enabled Wi-Fi communications from the perspective of energy efficiency. In particular, we are interested in investigating the Wi-Fi6 technology for addressing Wi-Fi-enabled IoT communications. Incorporating Wi-Fi chips into IoT devices can lead to a significant increase in power consumption. Therefore, we review the possible solution for this issue. Wi-Fi technology, with its high data rate standard like 802.11ac, has opened multiple opportunities for future wireless communications but became restricted for IoT communications due to its higher energy consumption when applied to battery operated IoT devices. Therefore, we have explored the power management perspective for Wi-Fi-enabled IoT communications. We have presented some simulation studies for the insight analysis of the issue. We validate our proposed model by considering varying traffic load, varying data rate, and varying number of IoT devices.

Keywords Internet of Things · Wi-Fi networks · IEEE 802.11ac · Low power consumption

B. P. Tewari (✉)
Department of Computer Science and Engineering, Ghani Khan Choudhury Institute of Engineering and Technology, Malda 732141, India
e-mail: babul@gkciet.ac.in

P. Mukherjee (✉)
Department of Computer Science and Engineering, National Institute of Technology, Durgapur 713209, India
e-mail: pm.20cs1105@phd.nitdgp.ac.in

A. Acharyya et al. (eds.), *Real-World Applications and Implementations of IoT*, Studies in Smart Technologies, https://doi.org/10.1007/978-981-97-8627-5_6

1 Introduction

The Internet of Things (IoT) extends the Internet communications among devices, everyday objects, and systems without human-to-human intervention. In recent times the use of IoT devices has been increased to a significant extent which causes a great flourish in IoT market [1]. IoT encompasses its wide applications in different domains such as household appliances [2], smart city [3], industrial machinery [4], health [5], and edge computing [6]. Many IoT devices operate on batteries. Therefore, energy efficiency is crucial to prolong battery life and minimize the need for frequent recharging or replacement. Devices that consume excessive power may not be practical for certain applications, especially those deployed in remote or inaccessible locations. However, power-hungry components like wireless communication modules, sensors, and processing units can drain batteries quickly. Maximizing battery life while maintaining functionality raises the need for energy-efficient IoT communications. Some low-power wireless technologies like ZigBee or Bluetooth low energy (BLE) proved its suitability for two IoT devices but they failed to support multiple IoT connections [1]. So, addressing the power consumption issue is crucial for the widespread adoption and sustainability of IoT technology.

Wireless local area network (WLAN) popularly known as Wi-Fi technology is a widely accepted as low-cost Internet service provider. This is due to its use in unlicensed spectrum. IEEE standard like IEEE 802.11ah has been considered as an effective means for IoT communications in unlicensed spectrum which can facilitate massive IoT communication [7]. Narrow-band Wi-Fi (NB-Wi-Fi) and its compatibility with the use of orthogonal frequency division multiplexing (OFDM), extends the capability of IEEE 802.11ax in IoT communications [8]. Unlike the high data rate applications at home like video streaming, online gaming with the use of Wi-Fi technology, the IoT applications may require low data rate. Furthermore, the power limitation is crucial for the battery-operated devices [9]. Keeping view to all these, appropriate Wi-Fi association for the IoT devices which will take care of simultaneous communications of high data rate Wi-Fi users and low data rate IoT devices with minimum power dissipation is an important concern of research.

The current manuscript addresses Wi-Fi-enabled IoT communications with the perspective of low power consumption. Wi-Fi-enable IoT can greatly reduce the traffic load in cellular communications. However, high data rate access of Wi-Fi service provider can lead to higher power consumption reducing the battery lifetime and enhancing the maintenance cost. So, the effective use of existing Wi-Fi infrastructure in implementing multiple IoT communications is a crucial challenge and needs a serious research attention. This paper focuses on the technical challenges of Wi-Fi-enabled IoT communications for implementing multiple IoT communications. It also describes an association model for IoT-enabled Wi-Fi communications. The association model is validated through some simulation results from the perspective of energy efficiency.

2 Relate Works

Coexistence of Wi-Fi and IoT communications is a praising research concern of recent times. In [1], the authors have addressed the technical challenges of multiple IoT connections under Wi-Fi networks and the possible solution of low-power IoT communications. In the same direction the role of IEEE 802.11ah standard has been described in [7]. The authors have described the support of massive IoT connections with 1 km range through IEEE 802.11ah networks with minimum access delay. The concept of NB-Wi-Fi has been demonstrated in [8] addressing the compatibility of OFDM with IEEE 802.11ax technology. In this paper, the authors advocate concurrent use of low-data rate IoT applications and high-data rate Wi-Fi technology under a common infrastructure. However, in NB-Wi-Fi massive IoT connections may increase the cost [1]. The power consumption of LTE-M and NB-Wi-Fi has been studied in [10]. The authors have reasonably addressed that size of traffic data does not influence much under suitable coverage however can lead to worse performance under the circumstance of the limited coverage. In [11], the authors have addressed an energy efficient IoT communications for WLANs. The key concepts they have introduced are to use a different carrier frequency of the AP's OFDM signal of IoT for data encoding and decoding at low sampling rate. This approach is cost-effective in comparison to the NB-Wi-Fi. The high data transmission of Wi-Fi causes a reasonably high energy consumption. To deal with this, the authors in [12] introduced a scheduling strategy using Wi-Fi and Zigbee in IoT communications. The objective is to reduce the energy consumption of IoT device. In [13], the increased contention of channel access for an increased number of regular Wi-Fi stations or IoT stations which in turn causes a high energy consumption has been addressed with downlink packet scheduling. The role of Wi-Fi6 in IoT environment has been addressed in [14]. The role of power consumption module of Wi-Fi in operating the IoT devices has been addressed in [15].

In Wi-Fi enabled IoT network, the selection of appropriate AP is also an influencing parameter from the perspective of energy efficiency. An energy efficient AP selection for IoT communications using reinforcement learning has been addressed in [16]. The authors emphasized on the issue of unbalanced load across the APs. A collection of several technologies in the context of wireless mesh networking with IoT has been discussed in [17]. The importance of appropriate AP selection for communicating IoT devices is discussed in [18]. The authors have considered strategic behavior of the objects and users.

Coexistence of IoT devices with Li-Fi users under a common Li-Fi AP has been addressed in [19]. The objective is to address the low power requirement of IoT devices with the power-hungry applications of Li-Fi users. An energy efficient AP selection scheme for heterogeneous networks consisting of Wi-Fi and Li-Fi technology has been proposed in [20]. They have considered both energy efficiency and QoS provisioning.

With the significant increase in the smart devices IoT has become an integral part of future wireless network [21, 22]. Battery life is crucial for the IoT devices [23]. In such context development of energy efficient IoT communications is a challenging concern [24]. This is because the lifetime of battery-operated devices is dependent on the power consumption [9].

Since the Wi-Fi network results in multiple transmission overlapping therefore association management of the devices plays a crucial role in determining the network performance [25, 26]. A frame aggregation-based association management has been addressed in [27]. However, such Wi-Fi associations [25–27] are suitable for high data rate applications and may result in greater power consumptions. This is unlike the IoT device requirements which are mostly battery operated and require low power association model [28]. Therefore, maintaining low power consumption remains a crucial design issue for Wi-Fi-enabled IoT communications [29]. In this context, power saving model of IEEE 802.11 standards has been considered as an effective means for energy saving associations [29, 30]. An energy aware task allocation strategy in IoT networks considering user mobility has been considered in [31]. User mobility in IoT has also been addressed in [32].

It is widely studied that Wi-Fi enabled IoT requires a careful design of association strategy. This is because IoT requires a low power association strategy, on the other hand traditional Wi-Fi devices are high throughput applications oriented [33] which may cause higher energy consumption. So appropriate association model in such heterogeneous environment is much essential and which is the main concern of this chapter.

3 Major Challenges and Motivations

In Wi-Fi enabled IoT communications, Wi-Fi network may experience congestion when the number of connecting IoT devices is increased. In such scenario, both the IoT devices and Wi-Fi devices access the network from the common service provider in unlicensed spectrum. Here, the major challenge is to ensure the different requirements of heterogeneous devices. IoT devices are based on low-rate applications expected to operate with low energy consumption. Wi-Fi devices are expected to meet high data rate applications like online video streaming or voluminous file downloading. A typical model of such Wi-Fi enabled IoT communications is shown in Fig. 1. In such scenario there are several technical challenges of different solutions to maintain a harmonic coexistence of these heterogeneous devices.

- **Limited in range**: Wi-Fi networks are limited by their coverage region. IoT devices sometimes require an extended connectivity and may be in places of difficult to reach.
- **Volume of aggregate data**: With the advancement of time the number of IoT connections may become huge. So individual data may be less however the aggregate data of the devices is substantial and can play a crucial role in performance. This should be addressed by suitable deployments of modern IEEE 802.11 standards like 802.11ax.

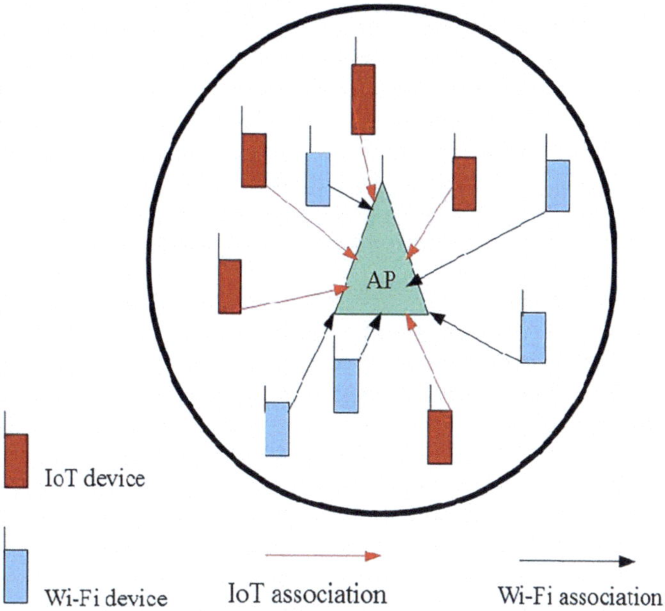

Fig. 1 IoT and Wi-Fi coexistence

- **Concurrent communications**: Ensuring a concurrent communication is a challenging task as there could be a huge difference in the required data rate between the IoT devices and Wi-Fi devices.
- **Power consumption**: IoT devices often operate on battery power, and Wi-Fi communication can be relatively power-intensive compared to other wireless technologies like Bluetooth Low Energy (BLE) or LoRaWAN. Minimizing power consumption while maintaining reliable communication is a significant challenge for IoT deployments in Wi-Fi networks.
- **Interference**: Due to the unlicensed nature of the spectrum, Wi-Fi networks can suffer from interference from other devices operating in the same frequency bands, such as Bluetooth devices, cordless phones, and microwave ovens. This interference can degrade the quality of IoT communications.

Addressing these challenges requires careful planning, deployment, and management of Wi-Fi networks for IoT applications, including optimization of network resources, implementation of interference mitigation techniques, and adoption of power-efficient communication protocols.

A time-scheduled IoT communication is often advocated as one of the feasible means to avoid the issue of high data rate Wi-Fi applications and low rate IoT applications. However, this approach will lead low spectrum efficiency for the Wi-Fi users and will not result in an effective means of communications. Thus, we advocate a concurrent communication which also ensures a reasonable power consumption strategy for the battery-operated IoT devices.

4 Wi-Fi Enabled IoT: A Low Power Consumption Model

Keeping a view of the low energy consumption requirements of the devices, we advocate for a low power association model for the IoT devices to access network services from Wi-Fi network. The objective of the proposed model is of two folds. First, to allow concurrent communication services of the two devices. Second, to ensure the high data rate applications of the Wi-Fi devices alongside a minimum power consumption of the IoT devices. The proposed model is shown in Fig. 2.

The model parameters are as follows:

- **Data rate**: The data rate of associating Wi-Fi device (WD_i) from AP_j denoted as R_{ij} is greater than certain threshold. So, $R_{ij} \geq R_{th}$.
- **Consumed energy**: The consumed energy k^{th} IoT device E_k is less than certain threshold E_{th}. So, $E_k \leq E_{th}$.

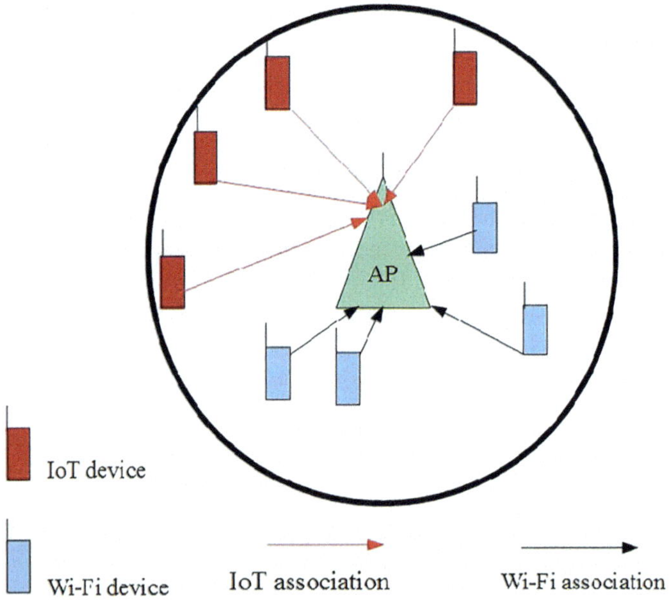

Fig. 2 Proposed association model

5 Protocol Validation

The proposed model is in consideration of the following protocol in use.

- An AP can provide service association to multiple IoT devices.
- A single OFDM carrier may be used by the IoT device while communicating with the AP.
- OFDM ensures multiple IoT connections by a single AP.

6 Performance Evaluation

The proposed model is evaluated through software simulation. We have considered a $100 \times 100\ m^2$ area where an AP is deployed at the central locations. The IoT devices are placed randomly and in the range where it can be served with a required data rate. The major simulation parameters used in this experimental set up are summarized in Table 1.

We have explored the performance evaluation from the perspective of consumed energy based on the following parameters:

- Varying data rate.
- Varying traffic size.
- Varying number of IoT devices.

6.1 Energy Consumption Versus Data Rate

Figure 3 represents the variations in energy consumption with the varying data rate. In this simulation, we have considered two variations in simulation times of 5 ms and 2.5 ms respectively. Reduction in simulation time causes less traffic load and hence for a fixed data rate we have a reduced energy consumption. Furthermore, it is interesting to note that the trend of average energy consumption for an increasing rate is decreasing which is quite expected. Additionally, it is very much noted that data

Table 1 Simulation parameters

Simulation parameters	Value
Wi-Fi range	100 m
Maximum number of IoT devices	10
Maximum data rate	18 Mbps
Simulation time	5 ms
AP power	20 dBm
Number of input task	1
Maximum input data size	15 Kb

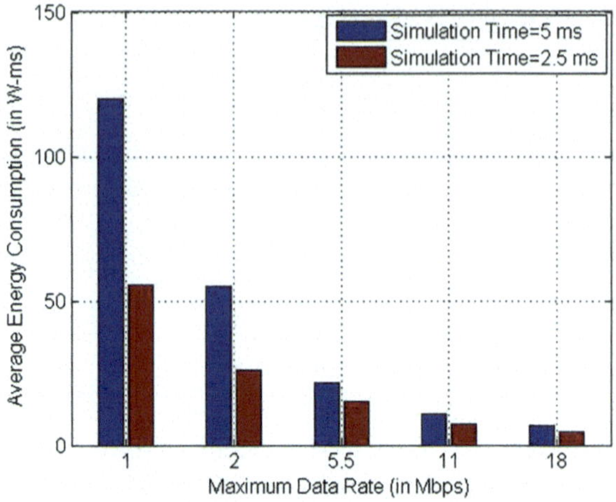

Fig. 3 Energy consumption for varying data rate

rate is an influencing parameter for the power dissipation of the IoT devices which is advocated by the proposed model. Note that in this simulation we have considered a single task to be offloaded by the IoT devices and the reference traffic size is 8 Kb.

6.2 Energy Consumption Versus Traffic Size

Traffic size plays a vital role in determining the energy consumption of the devices. This is because IoT devices are expected to operate at low rate and hence increased size leads to increased power dissipation which may endanger the battery life time. Figure 4 represents this. In such case, we have considered a fixed rate of 2 Mbps and vary the traffic load. We have considered two different variations of the simulation times.

As evident in Fig. 4, for a fixed data rate if we increase the size of the traffic it causes an increased power dissipation. Note that we have represented the average power consumption. Nevertheless the aggregate data traffic will be an influencing parameter for determining the battery lifetime as argued in proposed model.

6.3 Energy Consumption Versus Number of IoT Devices

It is argued in the proposed model that the expected data rate of the IoT devices is low. Furthermore, IoT devices are supposed to operate with low power consumption to increase the battery lifetime. However, the aggregate data may become voluminous when we consider all the IoT devices in operation. In such case total energy consumption is crucial.

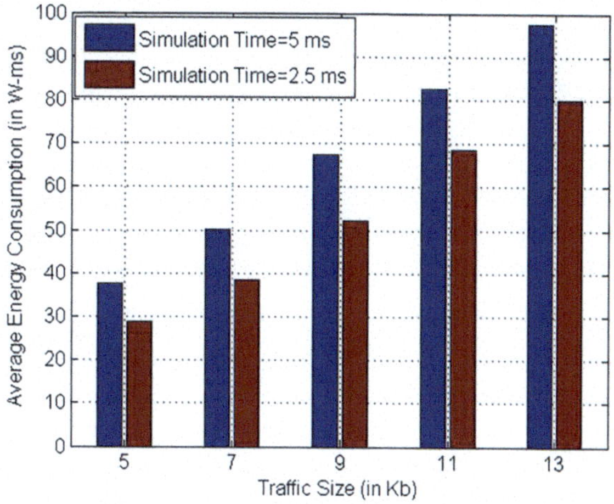

Fig. 4 Average energy consumption versus traffic size

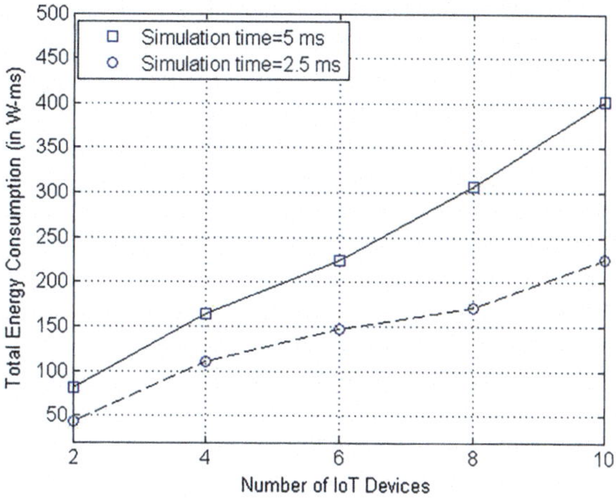

Fig. 5 Total energy consumption versus number of IoT devices

Figure 5 demonstrates how the total power consumption increases with the increase in the number of IoT devices. This is because of the increase in the aggregate data traffic. Again the obvious difference in terms of simulation time is noted. We argue, that aggregate data traffic is a valuable parameter for maintaining the balanced power consumption of the IoT devices.

7 Conclusion

This chapter addresses the role of Wi-Fi for IoT communication from the perspective of energy consumption. The importance of dissipated amount of power has been described with detailed insight information. In this context a model has been proposed considering the energy consumption metric to admit IoT communications under Wi-Fi networks. The objective is to address a harmonic coexistence of the IoT devices and Wi-Fi devices as well. Overall, optimizing energy consumption in IoT devices is crucial for ensuring sustainability, cost-effectiveness, reliability, and scalability of IoT deployments across various sectors. It involves the dsesign of energy-efficient hardware, software, and communication protocols, as well as the implementation of effective power management strategies throughout the lifecycle of IoT devices. Future work includes the appropriate placement of the IoT devices in Wi-Fi network.

References

1. Pirayesh H, Sangdeh PK, Zeng H (2020) Coexistence of Wi-Fi and IoT communications in WLANs. IEEE Internet Things J 7(8):7495–7505
2. Stojkoska BLR, Trivodaliev KV (2017) A review of internet of things for smart home: challenges and solutions. J Cleaner Prod 140:1454–1464
3. Samih H (2019) Smart cities and internet of things. J Inf Technol Case Appl Res 21(1):3–12
4. Salih KOM, Rashid TA, Radovanovic D, Bacanin N (2022) A comprehensive survey on the internet of things with the industrial marketplace. Sensors 22(730):2022
5. Alam MM, Malik H, Khan MI, Pardy T, Kuusik A, Le Moullec Y (2018) A survey on the roles of communication technologies in IoT-based personalized healthcare applications. IEEE Access 6:36611–36631
6. Bhatia M (2022) Energy efficient IoT-based informative analysis for edge computing environment. Trans Emerg Telecommun Technol 33(9)
7. Jamali A, Ghahfarokhi BS, Abedini M (2022) Improving performance of association control in IEEE 802.11ah-based massive IoT networks. IEEE Internet Things J 9(11):8572–8583
8. Wilhelmsson LR, Lopez MM, Sundman D (2017) NB-WiFi: IEEE 802.11 and bluetooth low energy combined for efficient support of IoT. In: 2017 IEEE wireless communications and networking conference (WCNC), San Francisco, CA, USA, pp 1–6
9. Mukherjee P, Tewari BP, De T (2023) Joint resource allocation and cluster-head selection for energy aware D2D multi-casting. Int J Ad Hoc Ubiquitous Comput 44(3):131–147
10. Labdaoui N, Nouvel F, Dutertre S (2023) Energy-efficient IoT communications: a comparative study of long-term evolution for machines (LTE-M) and narrowband internet of things (NB-IoT) technologies. In: IEEE symposium on computers and communications (ISCC), pp 823–830
11. Pirayesh H, Sangdeh PK, Zeng H (2019) EE-IoT: an energy-efficient IoT communication scheme for WLANs. In: IEEE INFOCOM 2019—IEEE conference on computer communications, pp 361–369
12. Malik A, Kushwah R (2023) Energy-efficient scheduling in IoT using Wi-Fi and ZigBee cross-technology. J Supercomput 79:10977–11006
13. Sheth J, Dezfouli B (2019) Enhancing the energy-efficiency and timeliness of IoT communication in WiFi networks. IEEE Internet Things J 6(5):9085–9097
14. János Varga P, Illési Z (2022) Wi-Fi 6 application in Iot environment interdisciplinary description of complex systems 20(3):277–283
15. Erkan Yüksel M (2020) Power consumption analysis of a Wi-Fi-based IoT device. Electrica 20(1):62–70
16. Lee S, Park J, Choi H, Oh H (2023) Energy-efficient AP selection using intelligent access point system to increase the lifespan of IoT devices. Sensors 23(5197):2023
17. Cilfone A, Davoli L, Belli L, Ferrari G (2019) Wireless mesh networking: an IoT-oriented perspective survey on relevant technologies. Futur Internet 11:99

18. Rezaeiye PP, Sharifi1 A, Rahmani AM, Dehghan M (2021) Access point selection in the network of internet of things (IoT) considering the strategic behavior of the things and users. J Supercomput 77:14207–14229
19. Anwar DN, Ahmad R, Srivastava A (2022) Energy-efficient coexistence of LiFi users and light enabled IoT devices. IEEE Trans Green Commun Netw 6(2):930–950
20. Asad M, Qaisar S (2022) Energy efficient QoS-based access point selection in hybrid WiFi and LiFi IoT networks. IEEE Trans Green Commun Netw 6(2):897–906
21. Nguyen DC et al (2022) 6G internet of things: a comprehensive survey. IEEE Internet Things J 9(1):359–383
22. Chettri L (2020) A comprehensive survey on internet of things (IoT) toward 5G wireless systems. IEEE Internet Things J 7(1):16–32
23. Jolly B (2021) IoT device battery life: go slow for fast insights into challenging conditions. In: 2021 IEEE international midwest symposium on circuits and systems (MWSCAS), pp 680–683
24. Rani S, Talwar R, Malhotra J, Ahmed SH, Sarkar M, Song H (2015) A novel scheme for an energy efficient internet of things based on wireless sensor networks. Sensors 15:28603–28626
25. Tewari BP, Ghosh SC (2019) Interference aware frequency assignment and association control for uplink and downlink traffic in WLAN. Int J Commun Netw Distrib Syst 23(2):143–171
26. Tewari BP, Ghosh SC (2017) Joint frequency assignment and association control to maximize the aggregate throughput in IEEE 802.11 WLAN. Wirel Pers Commun 94(3):1193–1221
27. Tewari BP (2023) Frame aggregation aware cluster based association management for heterogeneous users. J Ambient Intell Humaniz Comput 14(3):2811–2826
28. Lee S, Choi H, Kim T, Park H-S, Choi JK (2022) A novel energy-conscious access point (eAP) system with cross-layer design in Wi-Fi networks for reliable IoT services. IEEE Access 10:61228–61248
29. Ramanna VK, Sheth J, Liu S, Dezfouli B (2021) Towards understanding and enhancing association and long sleep in low-power WiFi IoT systems. IEEE Trans Green Commun Netw 5(4):1833–1845
30. Liu S, Ramanna VK, Dezfouli B (2020) Empirical study and enhancement of association and long sleep in 802.11 IoT systems. In: GLOBECOM 2020—2020 IEEE global communications conference, pp 1–7
31. Yao J, Ansari N (2019) Energy-aware task allocation for mobile IoT by online reinforcement learning. In: ICC IEEE international conference on communications (ICC), pp 1–6
32. Xu L, Zhou X, Li Y, Cai F, Yu X, Kumar N (2022) Intelligent power allocation algorithm for energy-efficient mobile internet of things (IoT) networks. IEEE Trans Green Commun Netw 6(2):766–775
33. Wu B, Funabiki N, Roy SC, Rahman MM, Kong D, Fang S (2024) An application of throughput request satisfaction method for maximizing concurrent throughput in WLAN for IoT application system. Sensors 24:2173

Chapter 7
A Drone-Based System for Detecting Forest Medicinal Plants

Aritrya Chatterjee, Avirup Manjumder, Pratik Acharjee, and Jayabrata Goswami

Abstract Agriculture is one of the keys to the significant development of human civilization. In this dynamic world, automation has started its amazingness in each field of work. But automation within the field of agribusiness isn't as progressed comparatively. In this way, the requirement of creating an automation framework within the field of agriculture is inescapable. Where it is obliged to require care of the crops additionally avoid them from procuring any genuine infections, our project tends to make an independent drone for the discovery of infections at the early stages of edit development of medicinal plants in different forest area. This paper proposes an unused approach of innovation in agrarian field, i.e. rural drone which works on observing the medicinal plants image, structure, and properties. The design drone is developed by hardware and software system using Arduino UNO and Tinker CAD tools.

Keywords Drone · Automation · Arduino UNO · Tinker CAD

1 Introduction

The Earth's forests are among the most bio-diverse and valuable ecosystems, providing essential ecosystem services, including carbon sequestration, soil stabilization, and water regulation. Among the myriad species that inhabit these forests, medicinal plants hold a special significance. For centuries, these plants have been used by indigenous communities and traditional healers for their therapeutic properties, contributing significantly to global healthcare systems [1–5].

However, the sustainable management of forest medicinal plants is increasingly challenging due to various anthropogenic pressures, including deforestation, habitat

A. Chatterjee · A. Manjumder · P. Acharjee
University of Engineering and Management, New Town, Kolkata, India

J. Goswami (✉)
Netaji Subhas Open University, Saltlake, Kolkata, India
e-mail: goswamijayabrata@gmail.com

A. Acharyya et al. (eds.), *Real-World Applications and Implementations of IoT*, Studies in Smart Technologies, https://doi.org/10.1007/978-981-97-8627-5_7

fragmentation, climate change, and overexploitation. Monitoring and conserving these plants are thus critical for maintaining biodiversity and ecosystem health, as well as for preserving traditional knowledge and supporting local livelihoods.

Traditional methods of monitoring and inventorying forest medicinal plants are often labor-intensive, time-consuming, and limited in spatial coverage. These methods involve field surveys, which can be costly and logistically challenging, especially in remote or inaccessible areas. Moreover, they may not provide a comprehensive view of the plant species present in a given area, leading to gaps in our understanding of forest biodiversity and medicinal plant distribution.

In recent years, advances in drone technology have revolutionized the field of biodiversity monitoring and conservation. Drones, also known as unmanned aerial vehicles (UAVs) [6–8], offer a non-invasive, cost-effective, and efficient means of collecting high-resolution aerial imagery and data from forest ecosystems. Equipped with various sensors, including cameras, multispectral and hyperspectral sensors, LiDAR (Light Detection and Ranging), and GPS (Global Positioning System) receivers, drones can capture detailed information about the forest canopy, including the distribution and abundance of medicinal plants.

The use of drones in detecting forest medicinal plants offers several advantages over traditional methods. Drones can access remote or inaccessible areas, such as dense forests or steep terrain, where traditional surveys are impractical or dangerous. They can also cover large areas in a relatively short amount of time, providing a more comprehensive view of the forest landscape. Additionally, drones can capture data with high spatial and temporal resolution, allowing researchers to monitor changes in plant populations over time and assess the impact of environmental factors, such as climate change and land use change.

In this paper, we present a comprehensive review of the current state of drone technology in detecting forest medicinal plants. We discuss the various sensors and imaging techniques used in drone-based plant detection, highlighting their strengths and limitations. We also review the existing literature on the application of drones in biodiversity monitoring and conservation, focusing on their use in mapping medicinal plant species.

Furthermore, we discuss the challenges and opportunities associated with the use of drones in detecting forest medicinal plants, including regulatory constraints, data processing and analysis, and the integration of traditional knowledge with modern technology. We also propose future research directions and potential applications of drone technology in supporting the sustainable management of forest medicinal plants and biodiversity conservation.

Overall, we aim to demonstrate the potential of drone technology as a powerful tool for advancing our understanding of forest ecosystems and supporting the sustainable management of forest medicinal plants. By combining the capabilities of drones with innovative research methods, we can enhance our ability to conserve biodiversity, protect natural resources, and promote the sustainable use of forest ecosystems for future generations.

2 Model Analysis

A. Challenges in Medicinal Plant Identification

Traditional methods of identifying medicinal plants are often laborious and time-consuming. They typically involve field surveys conducted by botanical experts who must visually inspect and classify plant species based on their morphological characteristics. This process is not only resource-intensive but also reliant on the expertise of a limited number of individuals. As a result, the identification of medicinal plants in remote forest areas can be hindered by a lack of trained personnel and the high costs associated with conducting field surveys (Fig. 1).

Furthermore, the accurate identification of medicinal plants is crucial for ensuring their safe and effective use in traditional medicine. Misidentification or confusion with similar-looking plant species can lead to serious health risks, including poisoning or ineffective treatment. Therefore, there is a pressing need for more efficient and reliable methods of identifying medicinal plants, particularly in remote forest areas where access to botanical expertise is limited [9–12].

B. Conservation Challenges

In addition to identification challenges, the conservation of medicinal plants is also a pressing concern. Many medicinal plant species are under threat due to habitat destruction, deforestation, and overexploitation. The loss of these species could have significant implications for global health, as many traditional medicines are derived from plant-based sources.

Conservation efforts aimed at protecting medicinal plant species face numerous challenges, including limited resources, competing land-use priorities, and a lack of

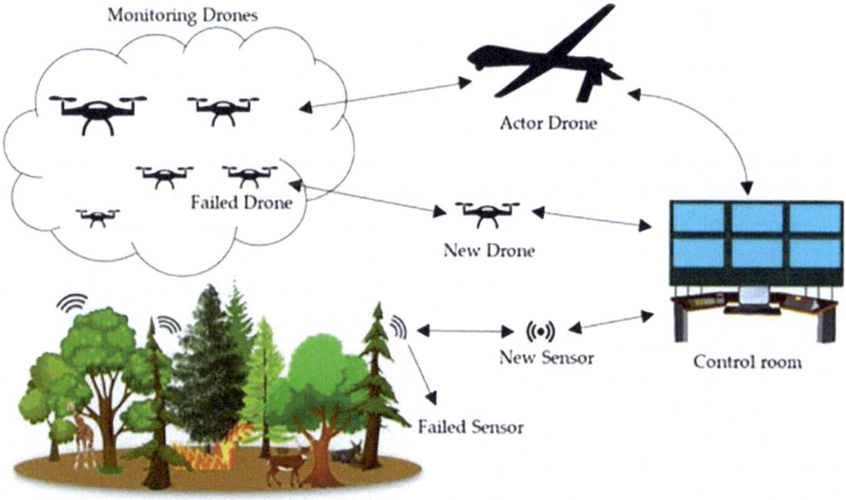

Fig. 1 Basic design model of forest medicinal plants detection system

awareness about the importance of these plants. Additionally, the sustainable use of medicinal plants presents a complex dilemma, as demand for these plants continues to rise while their natural habitats are increasingly threatened.

C. Innovative Solutions

Addressing the challenges of identifying and conserving medicinal plants requires innovative approaches that leverage technology and interdisciplinary collaboration. One such approach is the use of drones equipped with advanced imaging technology to survey remote forest areas and identify medicinal plant species.

Drones can cover large areas of forest quickly and efficiently, collecting high-resolution images that can be analyzed using machine learning algorithms to identify plant species. This approach has the potential to revolutionize the field of botanical surveying, enabling researchers to identify medicinal plants more quickly and accurately than ever before.

Another innovative solution is the development of portable DNA sequencing technologies that can rapidly identify plant species based on their genetic information. These technologies, often referred to as "DNA barcoding," allow researchers to identify plant species using small tissue samples, eliminating the need for visual inspection by botanical experts.

3 Design Model Structure and Operation

The proposed system utilizes drones with high-resolution cameras and GPS modules to capture images of forest areas. These images are then processed using machine learning algorithms to identify medicinal plants based on their visual characteristics. The system's design includes the following components:

(i) Drone Platform: The drone serves as the aerial platform for capturing images of forest areas. It is equipped with a high-resolution camera and a GPS module for accurate positioning.
(ii) Image Acquisition: The drone flies over the forest area, capturing images at regular intervals. The images are geotagged with their GPS coordinates for later analysis.
(iii) Image Processing: The captured images are processed using machine learning algorithms to identify medicinal plants. These algorithms are trained on a dataset of plant images to recognize specific plant species based on their visual features.
(iv) Data Analysis and Visualization: The identified plant species are displayed on a user-friendly interface, allowing conservationists and researchers to visualize the distribution of medicinal plants in the forest (Fig. 2).

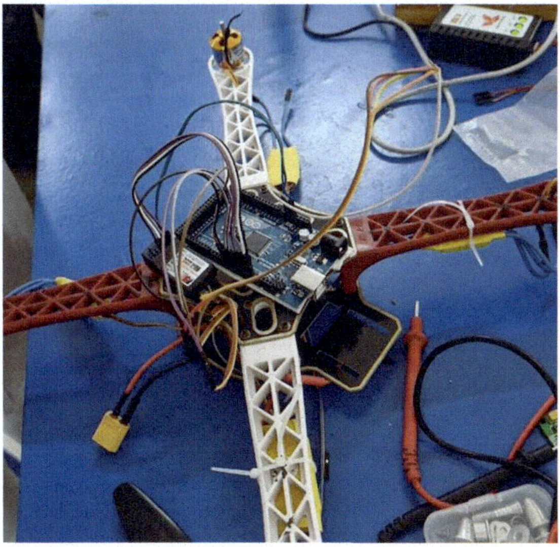

Fig. 2 Shows the basic model design of Arduino-UNO based drone

3.1 Implementation Details the System's Implementation Involves the Following Steps

(i) Dataset Preparation: A dataset of plant images is collected and labeled with the corresponding plant species.
(ii) Algorithm Training: Machine learning algorithms, such as convolutional neural networks (CNNs), are trained on the dataset to recognize medicinal plants.
(iii) Drone Flight: The drone is deployed to fly over the forest area, capturing images at regular intervals.
(iv) Image Processing: The captured images are processed using the trained algorithms to identify medicinal plants.
(v) Data Analysis: The identified plant species are analyzed to determine their distribution and abundance in the forest.
(vi) Data Transmission: The drones are equipped with high-speed data transmission systems that allow them to send the captured images and plant identification data to a ground station in real-time (Fig. 3).

4 Ground Station

The ground station is set up in a central location within the forest. It consists of computers running specialized software for receiving, processing, and analyzing the data transmitted by the drones. Operators can monitor the drones' progress, adjust their flight paths if needed, and view the results of the plant detection process.

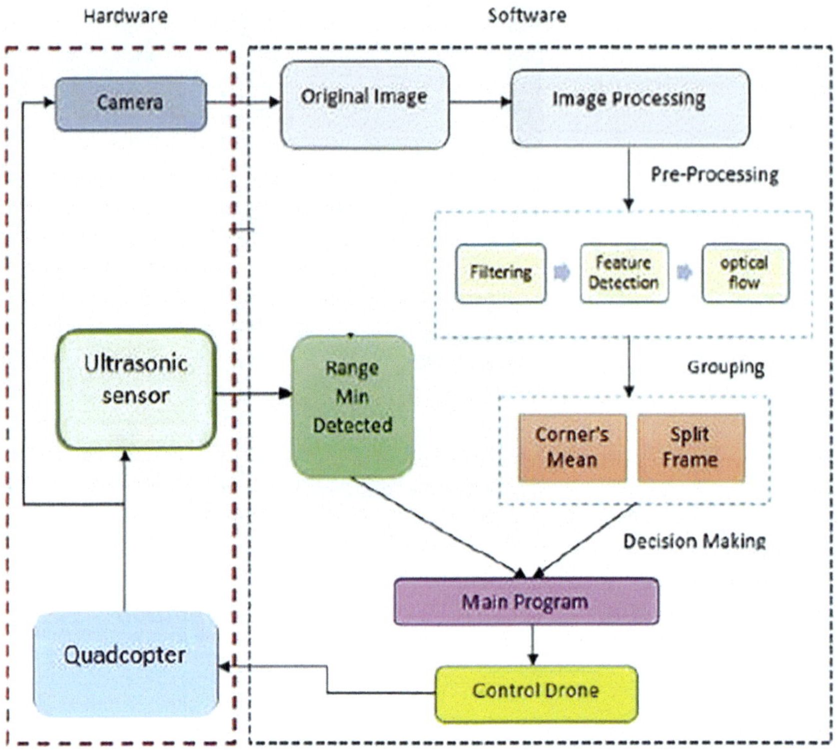

Fig. 3 Basic steps of design drone working model

5 Mapping Software

The software used in the ground station generates maps of the forest area showing the locations of the identified medicinal plants. These maps can be used to study the distribution patterns of the plants and plan conservation efforts.

6 User Interface

The ground station software includes a user-friendly interface that allows operators to easily control the drones, monitor their progress, and view the plant detection results. The interface provides real-time feedback and allows for quick decision-making in the field.

7 Power Systems

The drones are powered by high-capacity batteries that provide enough power for extended flight times, allowing them to cover large areas of the forest in a single mission.

8 Communication Systems

The drones are equipped with long-range communication systems that allow them to maintain contact with the ground station even when flying at a distance. This ensures reliable data transmission and control of the drones.

9 Environmental Sensors

In addition to cameras, the drones may also be equipped with environmental sensors to collect data on factors such as temperature, humidity, and soil conditions. This data can be used to study the impact of environmental factors on the distribution of medicinal plants.

Detecting medicinal plants through the use of drones involves several key operations, each serving a specific purpose in the process. By employing these various drone operations, researchers and conservationists can effectively detect and monitor medicinal plants in remote and inaccessible areas, aiding in their conservation and sustainable use. Here are various drone operations commonly used for detecting medicinal plants [13, 14]:

a. Aerial Surveys: Drones are used to conduct aerial surveys of forested areas to identify potential locations of medicinal plants. This operation involves flying the drone over the target area while capturing high-resolution images or videos of the vegetation below.
b. Multispectral Imaging: Drones equipped with multispectral cameras can capture images of plants at different wavelengths of light. This allows for the detection of specific plant characteristics, such as chlorophyll content, which can indicate the presence of medicinal compounds.
c. Hyperspectral Imaging: Similar to multispectral imaging, hyperspectral imaging involves capturing images of plants at hundreds of narrow and contiguous spectral bands. This provides detailed information about the biochemical composition of plants, aiding in the identification of medicinal species.
d. Data Analysis: Once the images are captured, they are processed using specialized software to identify medicinal plants based on their visual characteristics. This may involve image classification algorithms trained to recognize specific plant species.

e. Geotagging: Geotagging involves adding geographical coordinates to the images captured by the drone. This helps in creating accurate maps of the distribution of medicinal plants within the surveyed area.
f. Real-Time Monitoring: Drones can be used for real-time monitoring of medicinal plant populations, allowing researchers to track changes in plant distribution and abundance over time.
g. Precision Agriculture Techniques: Drones can also be used to apply precision agriculture techniques to medicinal plant cultivation. This includes the use of drones for seeding, spraying, and monitoring plant health.
h. Environmental Monitoring: In addition to detecting medicinal plants, drones can also be used to monitor environmental factors that affect plant growth, such as soil moisture, temperature, and light levels.

The above mentioned Fig. 4a, b represent the complete hardware and software model of the design of drone. The hardware model is implemented by Ardino-UNO and the software model is designed by Tinker CAD software tool (Table 1).

10 Experimental Results

Field Testing

The forest medicinal plants detection system using drones was tested in a forest area with a high density of medicinal plants. The drone was flown over the forest area, and the images and data collected by the drone were analyzed using the detection algorithm. The results showed that the system could detect medicinal plants with high accuracy.

Therefore Fig. 5 shows the complete working flow chart of medicinal plants detection system model work and results. So the system could also monitor the health and growth of the medicinal plants over time.

Figure 6a, b show the medicinal plants image, i.e. holy basil and turmeric plant which is very essential for the production of herbal medicine for human life which is captured by design drone model. The leaves of both plants are characterized by image detection and monitoring system by Arduino Uno. Python programming is used here for coding purpose.

11 Future Work

The forest medicinal plants detection system using drones can be further improved by integrating advanced image processing techniques and machine learning algorithms. The system can also be integrated with other technologies, such as GIS and remote sensing, to provide more comprehensive data on the distribution and abundance of

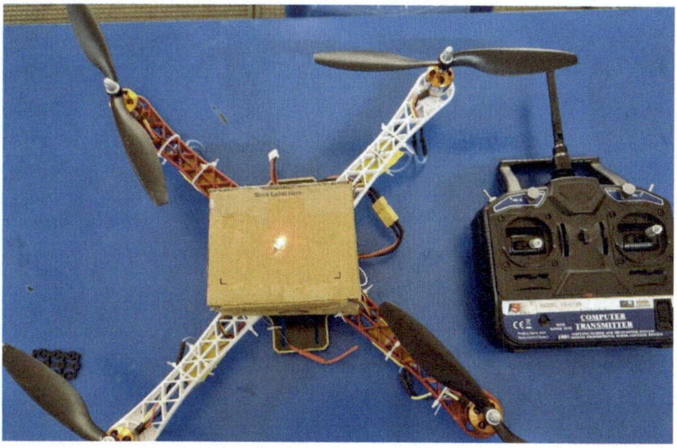

(a)

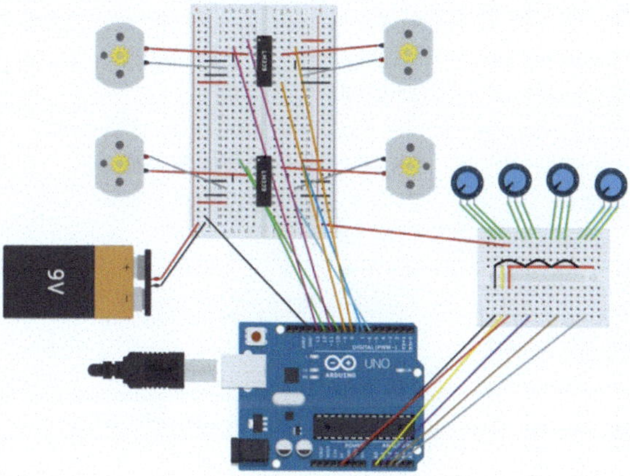

(b)

Fig. 4 **a** Hardware complete model of drone using Arduino Uno with wireless remote control. **b** Software model of drone using Tinker CAD software tool

medicinal plants. The system can also be used for the monitoring of other forest resources, such as timber and wildlife.

Table 1 Components required for hardware model

SL No	Device parameters name	Specification	Quantity
1	Quadcopter frame kit	DJI F450	1
2	Brushless motor	A2212 1000 kV	4
3	Brushless ESC	SimonK 30 A 2–3 S	4
4	8045 propeller	8045 carbon nylon propeller	4
5	1045 propeller	1045 propeller 10 in 10 × 4.5	4
6	Arduino Uno	Arduino Uno R3 compatible board + cable	1
7	Arduino nano	Arduino nano R3 compatible board with CH340 chip	1
8	MPU-6050	MPU-6050 6DOF 3 Axis Gyro with accelerometer sensor	1
9	LiPo battery	ABSD LiPo battery 11.1 V 2200 MAH 3S 80C	4
10	Strap belt	30 cm nylon strap belt for RC Lipo battery	4
11	XT60 male connector	XT60 male connector with 14AWG silicon wire 10 cm	4
12	FlySky FS-i6 transmitter	FlySky FS-i6 2.4G 6CH AFHDS RC transmitter with FS-iA6 receiver	1
13	Jumper cables	Combo of 3 type jumper cables	3
14	Ultra sonic sensor	HC-SR04	1

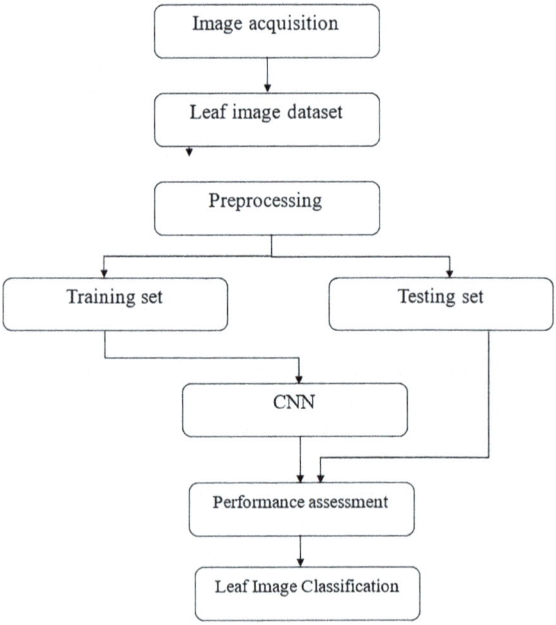

Fig. 5 Shows the working flow chart of the drone model system

(a) (b)

Fig. 6 **a** Holy basil plant image and **b** turmeric plant image taken by design drone

12 Conclusion

The forest medicinal plants detection system using drones is a novel approach for the detection and monitoring of medicinal plants in forests. The system can help in the identification and monitoring of medicinal plants in a non-destructive and efficient manner. The system can also provide valuable data for the conservation and management of medicinal plants. The system has the potential to revolutionize the way medicinal plants are identified and collected, and it can contribute to the sustainable use of forest resources.

References

1. Alsalam BHY, Morton K, Campbell D, Gonzalez F (2017) Autonomous UAV with vision based on-board decision making for remote sensing and precision agriculture. In: 2017 IEEE aerospace conference, Big Sky, M T. IEEE, pp 1–12. https://doi.org/10.1109/AERO.2017.7943593
2. Alvarez-Taboada F, Paredes C, Julián-Pelaz J (2017) Mapping of the invasive species Hakea sericea using unmanned aerial vehicle (UAV) and WorldView-2 imagery and an object-oriented approach. Remote Sens 9(9):913. https://doi.org/10.3390/rs9090913
3. Baron J, Hill D, Elmiligi H (2018) Combining image processing and machine learning to identify invasive plants in high-resolution images. Int J Remote Sens 39(15–16):5099–5118. https://doi.org/10.1080/01431161.2017.1420940
4. Belgiu M, Drăguţ L (2016) Random forest in remote sensing: a review of applications and future directions. ISPRS J Photogramm Remote Sens 114:24–31. https://doi.org/10.1016/j.isprsjprs.2016.01.011
5. Blaschke T (2010) Object based image analysis for remote sensing. ISPRS J Photogramm Remote Sens 65(1):2–16. https://doi.org/10.1016/j.isprsjprs.2009.06.004

6. Dvořák P, Müllerová J, Bartaloš T, Brůna J (2015) Unmanned aerial vehicles for alien plant species detection and monitoring. Int Arch Photogramm Remote Sens Spat Inf Sci 40(1):83–90. https://doi.org/10.5194/isprsarchives-XL-1-W4-83-2015
7. Göktogan A, Sukkarieh S, Bryson M, Randle J, Lupton T, Hung C (2010) A rotary-wing unmanned air vehicle for aquatic weed surveillance and management. J Intell Rob Syst 57(1):467–484. https://doi.org/10.1007/s10846-009-9371-5
8. Pajares G (2015) Overview and current status of remote sensing applications based on unmanned aerial vehicles (UAVs). Photogramm Eng Remote Sens 81(4):281–330. https://doi.org/10.14358/PERS.81.4.281
9. Ronneberger O, Fischer P, Brox T (2015) U-Net: convolutional networks for biomedical image segmentation. In: International conference on medical image computing and computer-assisted intervention, Munich, Germany. Springer, Cham, pp 234–241. https://doi.org/10.1007/978-3-319-24574-4_28
10. Kamilaris A, Prenafeta-Boldu FX (2018) Deep learning in agriculture: a survey. Comput Electron Agric 147:70–90. https://doi.org/10.1016/j.compag.2018.02.016
11. Martin F-M, Müllerová J, Borgniet L, Dommanget F, Breton V, Evette A (2018) Using single- and multi-date UAV and satellite imagery to accurately monitor invasive knotweed species. Remote Sens 10(10):1662. https://doi.org/10.3390/rs10101662
12. Van Wilgen B, Richardson D (1985) The effects of alien shrub invasions on vegetation structure and fire behaviour in South African fynbos shrublands: a simulation study. J Appl Ecol 22:955–966. https://doi.org/10.2307/2403243
13. Zhang L, Zhang L, Du B (2016) Deep learning for remote sensing data: a technical tutorial on the state of the art. IEEE Geosci Remote Sens Mag 4(2):22–40. https://doi.org/10.1109/MGRS.2016.2540798
14. Lehmann J, Prinz T, Ziller S, Thiele J, Heringer G, Meira-Neto J, Buttschardt T (2017) Open-source processing and analysis of aerial imagery acquired with a low-cost unmanned aerial system to support invasive plant management. Front Environ Sci 5:44. https://doi.org/10.3389/fenvs.2017.00044

Chapter 8
Skin Cancer Detection Using a Deep-Learning Based Framework

Mridul Ghosh, Arun Kumar Maiti, Priya Sarkar, Akash Jana, and Annapurna Roy

Abstract Skin cancer represents a prevalent cancer type, and early detection significantly impacts patient outcomes. A concerning rise in the incidence of various skin diseases is found nowadays. This alarming movement has not only affected the overall health and well-being of people but has also led to a notable rise in the development of skin cancer cases worldwide. This paper focuses on the development of a skin cancer detection system using a deep learning-based framework. It has been observed that deep learning outperforms conventional machine learning methods to a great extent. Among different deep learning methods, the Convolutional Neural Network (CNN) is very popular since it can handle spatial data well. The objective of this study is to leverage the performance of detecting skin cancer among the pool of skin lesions as benign or malignant by providing an automated tool to assist dermatologists in diagnosing the disease correctly in no time. Here, we have developed a CNN model to classify the skin lesions as benign or malignant. The performance of the model was evaluated using different performance measuring parameters such as accuracy, precision, recall, and F1-score. The key findings of this paper demonstrate the effectiveness of CNNs in skin cancer detection. The developed model achieved a satisfactory performance in classifying skin lesions, outperforming existing methods. The interpretability of the model was also explored, shedding light on the learned features and contributing to its potential as a diagnostic aid. Overall, this paper contributes to the advancement of computer-aided diagnosis systems for skin cancer. The developed CNN-based model shows promise in assisting dermatologists with accurate and efficient diagnoses.

M. Ghosh (✉) · A. Roy
Department of Computer Science, Shyampur Siddheswari Mahavidyalaya, Ajodhya, Shyampur, West Bengal 711312, India
e-mail: mridulxyz@gmail.com

A. K. Maiti
Department of Mathematics, Shyampur Siddheswari Mahavidyalaya, Ajodhya, Shyampur, West Bengal 711312, India

P. Sarkar · A. Jana
Department of Computer Science, Surendranath College, Kolkata, West Bengal 711312, India

A. Acharyya et al. (eds.), *Real-World Applications and Implementations of IoT*, Studies in Smart Technologies, https://doi.org/10.1007/978-981-97-8627-5_8

Keywords CNN · Skin cancer · Deep learning · Skin lesion · Classification

1 Introduction

Melanoma, the most lethal skin cancer type, presents a substantial global public health concern. It originates from the unchecked proliferation of melanocytes, the cells accountable for melanin pigment production in the skin. Without early detection and intervention, melanoma can metastasize, spreading to distant body areas, outcomes a grim prognosis and potentially life-threatening consequences. In other way, benign skin lesions, although not life threatening like melanoma, can still cause significant distress and discomfort to individuals. These benign lesions encompass a wide range of skin conditions, such as moles, cysts, lipomas, and seborrheic keratoses. While they may not have the same malignant potential as melanoma, they can still impact a person's life for diagnosis and management. Accurate and timely differentiation between melanoma and benign skin lesions is crucial for appropriate clinical decision-making and optimal patient outcomes. Dermatologists rely on their expertise, experience, and visual inspection to make these critical distinctions. However, this process can be challenging, as certain benign lesions may resemble melanoma in appearance, leading to potential misdiagnosis and unnecessary surgical interventions. Conversely, overlooking a suspicious lesion as benign can result in delayed diagnosis and treatment, with adverse consequences for the patient. In light of these challenges, the development of a reliable tool that can differentiate between melanoma and benign skin lesions is of paramount importance. The application of advanced technologies, such as CNNs [1], can be pivotal in addressing this challenge. By harnessing the capabilities of deep learning [2, 3], CNNs can effectively learn complex patterns and features from extensive datasets of annotated skin images. These learned patterns can then be utilized to make accurate predictions about the nature of skin lesions, distinguishing between melanoma and benign conditions. Implementing such a CNN-based model [4] as a diagnostic tool is able to improve the performance of skin cancer prediction. The main aim of this research is to create and assess a CNN model designed for the automated identification and categorization of melanoma and benign skin lesions. Through the utilization of a varied dataset comprising annotated skin images for training purposes, our goal is to leverage deep learning techniques to achieve precise and impartial differentiation between these distinct categories. The outcomes of this research have the potential to revolutionize the field of dermatology and skin cancer diagnosis. By providing a reliable and efficient tool for prompt detection of melanoma and differentiation from benign lesions, healthcare professionals can intervene promptly, improving patient outcomes and potentially saving lives. Moreover, the utilization of such a model can help to reduce unnecessary surgical procedures. However, visual diagnosis of skin cancer poses challenges due to its subjective nature and the expertise required for accurate interpretation. Traditional diagnostic approaches heavily rely on the skills and experience of dermatologists, resulting in variations in diagnosis and potential delays in treatment. Moreover, the increasing burden of skin cancer cases places a strain on healthcare systems, further underscoring the need for efficient and accurate diagnostic tools. In light of these challenges, the motivation behind this paper is to leverage the power of deep learning, specifically CNNs [5], to develop an automated skin cancer detection system. CNNs [6] have demonstrated remarkable capabilities in image analysis tasks, excelling in feature extraction and pattern recognition. By training a CNN [7] model on a large dataset of annotated skin images, we aim to create a system that can effectively classify skin lesions as benign or malignant, providing an objective and reliable diagnostic tool. This automated method holds promise for enhancing diagnostic precision, minimizing subjectivity, and aiding healthcare professionals in making prompt and well-informed choices. Additionally, an automated system for detecting skin cancer could potentially improve diagnostic accessibility, especially in areas with restricted healthcare resources or underserved communities. By leveraging advancements in deep learning [8] and computer vision, we strive to develop an innovative solution that can aid in prompt detection of skin cancer, ultimately leading to improved patient outcomes and a positive impact on public health. This paper seeks to advance the field of medical image analysis and promote the development of state-of-the-art technologies for enhancing skin cancer diagnosis. The study aims to further the progress of medical image analysis and the utilization of CNNs [9] in diagnosing skin cancer accurately. By addressing the crucial requirement for precise discrimination between melanoma and benign skin lesions, this research holds the potential to substantially influence patient care, enhancing early detection rates and mitigating the morbidity and mortality linked with melanoma (Fig. 1).

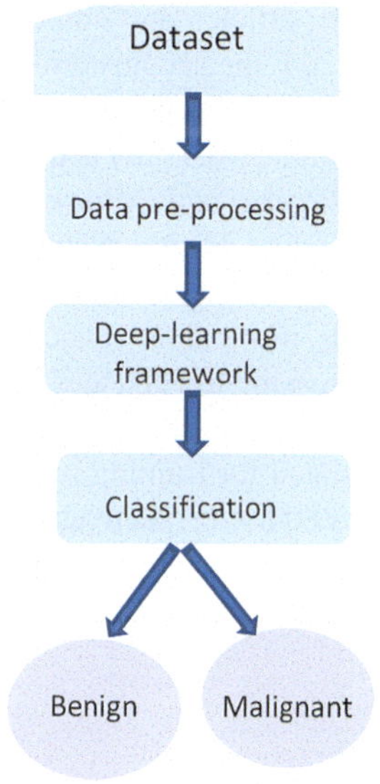

Fig. 1 Workflow diagram of our proposed approach

2 Literature Study

Hosny et al. [10] developed a skin lesion classification method using pre-trained deep learning and transfer learning to classify melanoma, common nevus, and atypical nevus. Dildar et al. [11] conducted a comprehensive review of deep learning techniques for skin cancer detection, with a specific focus on neural networks used in classifying lesion images. Hoshyar et al. [12] overviewed automatic skin cancer detection methods, emphasizing melanoma detection and automated diagnosis systems. Manne et al. [13] and Chang [14] discussed the evolution of CNNs in skin cancer classification, highlighting advantages and vulnerabilities compared to dermatologists.

Nahata and Singh [15] introduced a technique to classify melanoma and non-melanoma skin cancers with a high accuracy rate using the InceptionResnet model. Abdar et al. [16] described a hybrid dynamic method for derma cancer classification considering uncertainty, achieving high accuracy and F1-score. Dubal et al. [17] presented a method to detect and classify skin lesions as benign or malignant using general camera images, achieving 76.9% accuracy. Bhatt et al. [18] explored different techniques like k-nearest neighbors, support vector machines, and CNNs for this problem. Alquran et al. [19] used Dermoscopy image databases, in their work and exploited SVM and obtained 92.1% classification accuracy for melanoma detection. Vidya and Karki [20] used ABCD rule, GLCM, HOG, and classification using SVM, KNN, and Naïve Bayes. Achieved 97.8% accuracy and 0.94 AUC using SVM on 328 benign and 672 melanoma images, with 86.2% sensitivity and 85% specificity using KNN. Monika et al. [21] used pre-processing techniques which involved removing hair particles and smoothing the image. Segmentation is performed using color-based k-means clustering. Feature extraction is done using ABCD and GLCM. The experiment is conducted on the ISIC 2019 dataset and Multi-class Support Vector Machine (MSVM) is used for classification with an accuracy of 96.2%.

Phalke and Mhaske [22] emphasized the utilization of various classifiers, including Neural Network and Support Vector Machine, for early detection and classification of Melanoma skin cancer. By accurately identifying the cancerous cells, medical professionals can take the necessary steps to provide timely treatment, increasing the chances of a successful recovery. Lau and Al-Jumaily [23] proposed a system that classifies skin cancer types. Images are enhanced and cancer cells are isolated for training two neural networks. The system achieved accuracy of 89.9% and 80.8% respectively on a database of dermoscopy and digital photos.

3 Proposed Method

Dataset: The dataset employed in this study comprises a balanced collection of images portraying both benign and malignant skin moles. It consists of two folders, each containing 1800 pictures sized at 224 × 244 pixels, representing the two respective types of moles. This dataset was sourced from the ISIC-Archive of the International Skin Imaging Collaboration (ISIC) [24]. The primary objective of this research is to develop a model capable of visually classifying skin moles as either benign or malignant. The dataset encompasses two distinct classes of skin cancer: Benign and Malignant. Provided below (Fig. 2) are samples of input images.

Pre-processing: The images are rescaled to a range of 0 to 1 by setting the rescale parameter to 1/255, thereby normalizing pixel values for improved training efficacy. Subsequently, the dataset is partitioned into training and validation subsets to evaluate model performance on unseen data. Uniform resizing to dimensions (224, 224) is applied to ensure consistency in input size, facilitating standardized input dimensions for the model. These preprocessing steps ensure that input images are appropriately rescaled and resized prior to their incorporation into the training and evaluation phases.

CNN Components:

1. Convolutional Layers: Utilize filters to extract local patterns and features from input images, generating feature maps through convolution operations.
2. Activation Function: Introduces non-linearity post-convolution, commonly ReLU, facilitating the learning of complex relationships within the data. The Rectified Linear Unit (ReLU) activation function is defined as:

$$\mathrm{ReLU}(x) = \max(0,\ x) \tag{1}$$

   In simpler terms, if the input value (x) is greater than zero, ReLU returns the input value itself. Otherwise, it returns zero (Fig. 3).
3. Pooling Layers: Pooling layers reduce the spatial dimensions of feature maps while preserving important information, commonly achieved through max

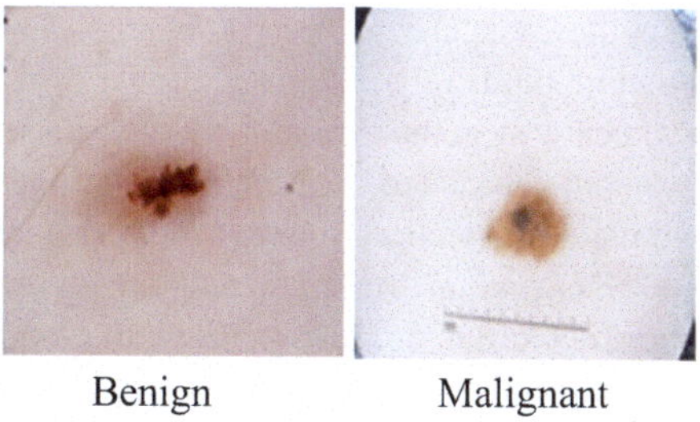

Fig. 2 Samples of input images

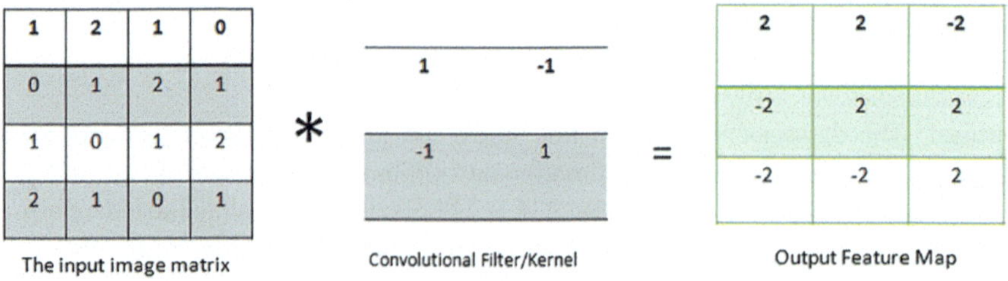

Fig. 3 An example on how convolutional works

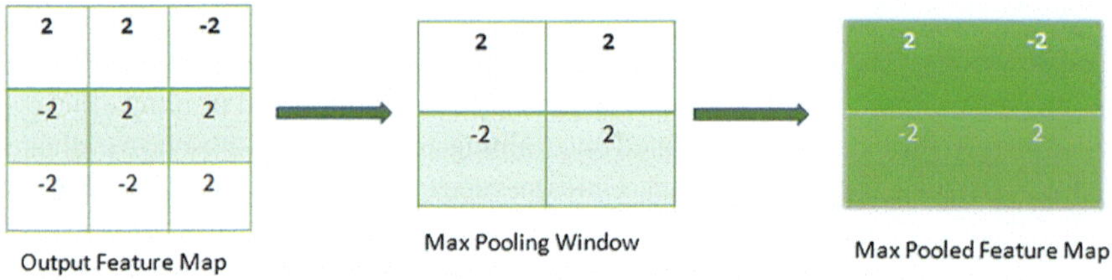

Fig. 4 An example on how max pooling works

pooling. Max pooling decreases the feature map's size, creating a downsampled representation. For instance, in Fig. 4, a 3×3 feature map is reduced to a 2×2 max-pooled feature map.

4. Fully Connected Layers: Flatten feature maps into 1D vectors, allowing traditional neural network-like processing to extract high-level features for predictions.
5. Output Layer: Typically comprises neurons representing specific classes in classification tasks, with activation functions tailored to the task, like softmax for multi-class classification.

4 Proposed Architecture

Skin cancer rates are increasing, necessitating early detection for improved treatment outcomes. Visual detection is subjective and skill-dependent, leading to delays. Traditional methods rely on dermatologist expertise, causing discrepancies and strain on healthcare systems. To address this, deep learning, specifically CNNs, is used to create an automated skin cancer detection system. CNNs excel at feature extraction and pattern detection in image analysis tasks.

CNNs, pivotal in computer vision, revolutionize tasks such as image classification, object detection, and image segmentation. Modeled after the human visual system, they autonomously glean significant features from visual data. Leveraging spatial correlations in images, CNNs employ dedicated layers like convolutions, pooling, and fully connected layers (Fig. 5).

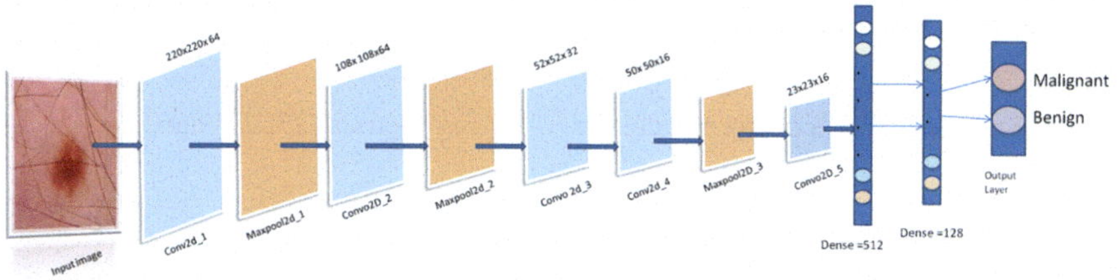

Fig. 5 Proposed (CP)2 C2PCD2 framework to classify Benign and Malignant skin cancer

We introduce a skin disease classification architecture termed (CP)2 C2PCD2, comprising 5 convolutional layers, 3 max-pooling layers, and 2 dense layers. The input image undergoes processing through various layers: initially, a convolutional layer (Conv2D) with 64 filters of size 5 × 5, followed by a max-pooling layer of 2 × 2. Subsequently, another Conv2D_1 with 64 filters of size 3 × 3 is applied, succeeded by a 2 × 2 max-pooling layer. Further, the Conv2D_2, Conv2D_3 with filters of 3 × 3 and 32, and another Conv2D_4 layer 3 × 3 and 16, respectively, are utilized to capture distinct features. Following each convolutional layer, a max-pooling layer of 3 × 3 is employed. A batch normalization layer enhances network stability, while dropout regularization with rates of 0.5 and 0.3 is applied after dense layers. The architecture includes two dense layers of sizes 512 and 128, respectively, performing weighted sums of input features. Finally, the output layer comprises 2 neurons representing classification categories. Detailed parameters are presented in Table 1.

Table 1 The number of parameters generated in our model

Layer (type)	Output shape	Parameter
Conv2D	(220, 220, 64)	4,864
Conv2D_1	(108, 108, 64)	36,928
Conv2D_2	(52, 52, 32)	18,464
Conv2D_3	(24, 24, 32)	4624
Conv2D_4	(22, 22, 16)	2320
Batch_normalizaion	8464	33,856
Dense	512	4,334,080
BatchNormalization-2	512	2,048
Dense-2	128	65,664
BatchNormalization-3	128	512
Dense-3	2	258
Total parameters	4,503,618	
Total trainable parameters	4,485,410	

5 Evaluation Protocol

To evaluate the model's performance, the following metrics [25–27] have been used in this work.

$$\text{Accuracy} = (\text{TP} + \text{TN}) / (\text{TP} + \text{TN} + \text{FN} + \text{FP}) \tag{2}$$

$$\text{Recall} = (\text{TP}) / (\text{TP} + \text{FN}) \tag{3}$$

$$\text{Precision} = (\text{TP}) / (\text{TP} + \text{FP}) \tag{4}$$

$$\text{F1 score} = \left(2^{*}\ \text{Precision}^{*}\ \text{Recall}\right) / (\text{Precision} + \text{Recall}) \tag{5}$$

5.1 Training Regime

Keeping the details of their respective labels (the type of mole), we load the photos from the two files into memory. To create training and test sets, we divided the dataset. Since the dataset has 1800 images per class, we choose a train-test ratio of 80:20. Therefore, the training set will receive 80% of the images for each class, while the test set will receive the remaining 20%. We also used a train-test ratio of 6:4 and 7:3. Before separating, we make sure to shuffle the data to remove any potential biases. The picture data is then normalized. To scale all pixel values to the range of 0 to 1, we divide each value by 255. This normalization process ensures that our neural network can effectively learn from the data.

6 Result and Discussions

We have trained the model through different epochs such as 50, 100, 150 and 200, and different train and test ratio.

6.1 Ablation Study

From Table 2 it is observed that C5P3D2 architecture gives us higher accuracy compared to the other combinations. The rest of the experimnet was performs on this C5P3D2 architecture. By varying the number of epochs the experiment was performed. From Table 3 it is observed that in 100 epochs the highest accuracy and precision are obtained. So we keep this epoch as constant in the rest of our experiment).

Based on the observations from Tables 3 and 4, it is noted that the highest accuracy and precision are achieved with an epoch setting of 100 and an 8:2 train-test ratio. Therefore, we maintain this ratio constant throughout the remainder of our experiment.

So, here it is noted that our proposed model achieved the highest accuracy for batch value 64 with train test ratio = 8:2 and over epoch = 100 (Table 5).

In Fig. 6 the training and validation loss and accuracy curves are shown.

Table 2 An ablation study conducted to develop a CNN-based framework aimed at achieving enhanced performance

Architecture	Convolutional	Pooling	Dense	Accuracy (%)
CPD	3 × 3	2 × 2	128	70.25
CCPDD	5 × 5, 3 × 3	3 × 3	256, 128	73.04
CCPD	5 × 5, 3 × 3	2 × 2	256	72.5
CPPD	5 × 5	3 × 3, 2 × 2	128	75.02
CCPPDD	5 × 5, 3 × 3	3 × 3, 3 × 3	256, 128	77.66
C5P3D2	5 × 5, 3 × 3, 3 × 3, 3 × 3, 3 × 3	2 × 2, 2 × 2, 3 × 3	512, 128	**80.42**

Table 3 Performance evaluation for different epochs using an 8:2 train-test ratio

Epoch	Accuracy (%)	Precision (%)	Recall (%)	F-score (%)
50	78.51	79.12	78.14	78.70
100	**78.76**	**79.02**	**78.76**	**78.52**
150	77.85	77.91	77.85	77.69
200	76.78	76.91	76.78	76.57

Table 4 Performance metrics for the epoch set at 100

Train-test	Accuracy (%)	Precision (%)	Recall (%)	F-score (%)
6:4	73.98	74.76	72.68	73.33
7:3	74.92	75.30	74.92	74.52
8:2	**78.76**	**79.02**	**78.76**	**78.52**

Table 5 Performance metrics for epoch set at 100 while varying batch sizes

Batch	Accuracy (%)	Precision (%)	Recall (%)	F-score (%)
32	78.76	79.02	78.76	78.52
64	**80.42**	**80.50**	**80.42**	**80.31**

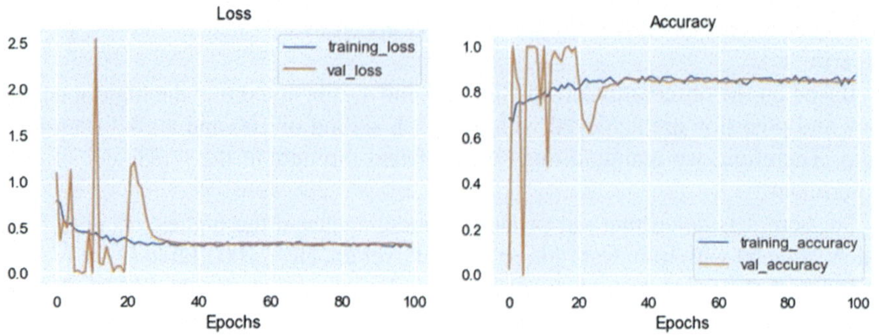

Fig. 6 Training and validation loss and accuracy curve for batch of 64, 8:2 train-test ratio, epoch of 100

Fig. 7 Confusion matrix

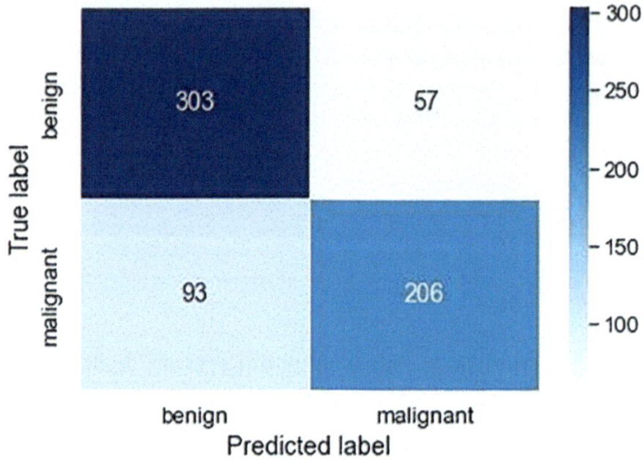

The confusion matrix for the accuracy of 80.42% for 64 batch sizes over 100 epochs is shown in Fig. 7.

Table 6 Comparison table with other authors

Author's name	Performance			
	Accuracy (%)	Precision (%)	Recall (%)	F-score (%)
Dubal et al. [17]	76.9	–	–	–
Alquran et al. [19]	92.1	–	–	–
Dildar et al. [11]	70.44	–	–	–
Liao et al. [28]	70.00	–	–	55.00
Proposed	80.42	80.50	**80.42**	80.31

True Positives (TP): 206 malignant samples were correctly identified as such by the model.

True Negatives (TN): 303 benign samples were accurately classified as benign by the model.

The model mistakenly identified 57 benign samples as cancer, leading to false positives (FP).

False Negatives (FN): The model predicted 93 benign samples as malignant in error.

Comparison Table: Table 6 showcases a comparison with other established techniques.

7 Conclusion

In summary, CNNs offer significant promise for skin cancer detection. Specifically designed for image analysis, CNNs excel in detecting and classifying skin cancer from images. Their advantages include automatic feature extraction, robustness with complex datasets, scalability, and potential for generalization to new cases. Furthermore, CNNs provide objective assessments, aiding dermatologists and healthcare professionals by enhancing diagnostic accuracy and reducing human error. Further research and development in the field of CNN-based skin cancer detection are necessary to refine and improve the models' performance, optimize their interpretability, and enhance their integration into clinical practice. Additionally, the deployment of CNNs in real-world healthcare settings requires addressing challenges such as data privacy, regulatory compliance, and the establishment of guidelines for their reliable and responsible use.

References

1. Ghosh M et al (2019) Identifying the presence of graphical texts in scene images using CNN. In: International conference on document analysis and recognition workshops (ICDARW), vol 1. IEEE
2. Ghosh M, Roy SS, Mukherjee H, Obaidullah SM, Santosh KC, Roy K (2022) Understanding movie poster: transfer-deep learning approach for graphic-rich text recognition. Vis Comput
3. Ghosh M, Mukherjee H, Obaidullah SM, Santosh KC, Das N, Roy K (2021) LWSINet: a deep learning-based approach towards video script identification. Multimed Tools Appl 80(19):29095–290128
4. Bappy JH, Roy-Chowdhury AK (2016) CNN based region proposals for efficient object detection. In: IEEE international conference on image processing (ICIP), pp 3658–3662
5. Ghosh M, Roy SS, Mukherjee H, Obaidullah SM, Gao XZ, Roy K (2021) Movie title extraction and script separation using shallow convolution neural network. IEEE Access 9:125184–125201
6. Zuluaga-Gomez J, Al Masry Z, Benaggoune K, Meraghni S, Zerhouni N (2021) A CNN-based methodology for breast cancer diagnosis using thermal images. Comput Methods Biomech Biomed Eng Imaging Vis 9(2):131–45
7. Ghosh M, Mukherjee H, Obaidullah SM, Roy K (2021) STDNet: a CNN-based approach to single-/mixed-script detection. Innovations Syst Softw Eng 17(3):277–288
8. Ghosh M, Obaidullah SM, Gherardini F, Zdimalova M (2021) Classification of geometric forms in mosaics using deep neural network. J Imaging 7(8):149
9. Obaidullah SM, Ghosh M, Mukherjee H, Roy K, Pal U (2022) SEN: stack ensemble shallow convolution neural network for signature-based writer identification. In: 26th International conference on pattern recognition (ICPR), pp 1414–1420
10. Hosny KM, Kassem MA, Foaud MM (2018) Skin cancer classification using deep learning and transfer learning. In: 9th Cairo international biomedical engineering conference (CIBEC). IEEE
11. Dildar M et al (2021) Skin cancer detection: a review using deep learning techniques. Int J Environ Res Public Health 18(10):5479
12. Hoshyar AN, Al-Jumaily A, Sulaiman R (2011) Review on automatic early skin cancer detection. In: International conference on computer science and service system (CSSS), p 40364039
13. Manne R, Kantheti S, Kantheti S (2020) Classification of skin cancer using deep learning, convolutional neural networks-opportunities and vulnerabilities-a systematic review. Int J Mod Trends Sci Technol ISSN 6:2455–3778
14. Chang H (2017) Skin cancer reorganization and classification with deep neural networks. arXiv: 1703.00534
15. Nahata H, Singh SP (2020) Deep learning solutions for skin cancer detection and diagnosis. In: Machine learning with health care perspective: machine learning and healthcare, pp 159–182
16. Abdar M, Samami M, Mahmoodabad SD, Doan T, Mazoure B, Hashemifesharaki R, Liu L, Khosravi A, Acharya UR, Makarenkov V, Nahavandi S (2021) Uncertainty quantification in skin cancer classification using three-way decision-based Bayesian deep learning. Comput Biol Med 135:104418
17. Dubal P, Bhatt S, Joglekar C, Patil S (2017) Skin cancer detection and classification. In: 6th international conference on electrical engineering and informatics (ICEEI), pp 1–6
18. Bhatt H, Shah V, Shah K, Shah R, Shah M (2022) State-of-the-art machine learning techniques for melanoma skin cancer detection and classification: a comprehensive review. Intell Med
19. Alquran H, Qasmieh IA, Alqudah AM, Alhammouri S, Alawneh E, Abughazaleh A, Hasayen F (2017) The melanoma skin cancer detection and classification using support vector machine. In: 2017 IEEE Jordan conference on applied electrical engineering and computing technologies (AEECT), pp 1–5

20. Vidya M, Karki MV (2020) Skin cancer detection using machine learning techniques. In: IEEE international conference on electronics, computing and communication technologies (CONECCT), pp 1–5
21. Monika MK, Vignesh NA, Kumari CU, Kumar MNVSS, Lydia EL (2020) Skin cancer detection and classification using machine learning. Mater Today Proc 33:4266–4270
22. Mhaske HR, Phalke DA (2013) Melanoma skin cancer detection and classification based on supervised and unsupervised learning. In: International conference on circuits, controls and communications (CCUBE), pp 1–5
23. Lau HT, Al-Jumaily A (2009) Automatically early detection of skin cancer: study based on neural network classification. In: International conference of soft computing and pattern recognition, pp 375–380
24. Bajwa MN et al (2020) Computer-aided diagnosis of skin diseases using deep neural networks. Appl Sci 10(7):2488
25. Ghosh M, Mukherjee H, Obaidullah SM et al (2023) Scene text understanding: recapitulating the past decade. Artif Intell Rev 56:15301–15373
26. Ghosh M, Roy SS, Roy B, Banik et al (2023) MOPO-HBT: a movie poster dataset for title extraction and recognition. Multimed Tools App
27. Ghosh M et al (2023) Plant disease detection and classification using a deep learning-based framework. In: Quaresma P, Camacho D, Yin H, Gonçalves T, Julian V, Tallón-Ballesteros AJ (eds) Intelligent data engineering and automated learning—IDEAL 2023, IDEAL 2023. Lecture notes in computer science, vol 14404. Springer, Cham
28. Liao H, Li Y, Luo J (2016) Skin disease classification versus skin lesion characterization: achieving robust diagnosis using multi-label deep neural networks. In: 23rd International conference on pattern recognition (ICPR), pp 355–360

Chapter 9
Stability Analysis of Injection Locked Optoelectronic Oscillator: A Fascinating Technology with Potential Applications in the Internet of Things (IoT)

Paromita Das and Dia Ghosh

Abstract In this literature, we explore the stability analysis of a time-delayed injection locked Single Loop Optoelectronic Oscillator (SLOEO). To this end, we develop analytically a delay dynamical system equation. We show for the first time that interplay between feedback loop delay, nonlinearity, injected signal strength leads to Hopf bifurcation and period doubling bifurcation in the system. We discuss the bifurcation diagrams in terms of the Injection strength and feedback loop delay. OEOs are an essential component of modern communication technology as it finds many important applications in Internet of Things (IoT). Therefore it is important to understand their collective dynamical behavior.

Keywords Optoelectronic oscillator · Delay dynamical system · Hopf bifurcation · Injection locking

1 Introduction

In the recent past, a revolution has been seen in communication technology due to a rapid growth of basic scientific discoveries. Nowadays optoelectronic devices and optical fiber communication technology are considered as the backbone of the communication engineering. Optoelectronic devices have various advantages over conventional RF practices, these include large bandwidth, high speed and efficiency, low noise, small footprints, and comparatively simple operation. A number of varied applications such as Doppler radars, satellite communication, metrology, or sensor which operate at ultra-high frequencies, require extremely stable oscillators competent enough to produce ultra-pure signals at the frequencies of tens of gigahertz and can be easily interfaced with optical systems. An Optoelectronic oscillator (OEO)

P. Das · D. Ghosh (✉)
Department of Electronics and Communication Engineering, Siliguri Institute of Technology, Siliguri, West Bengal 734009, India
e-mail: dia.slg42@gmail.com

A. Acharyya et al. (eds.), *Real-World Applications and Implementations of IoT*, Studies in Smart Technologies, https://doi.org/10.1007/978-981-97-8627-5_9

has excellent capability to meet this requisite. The stupendous performances of OEO instigate from the long optical fiber feedback loop. The delay line in the feedback loop produces many cavity modes with very narrow spacing. Out of these several RF mode frequencies, the RF filter allows only one mode to oscillate in a continuous and stable way. OEO is an intrinsic delay dynamical oscillator as it contains optical fiber delay line in the feedback loop. Delay dynamical systems are prolific area of research in different fields, which include physics, biology, mathematics, engineering, and ecology. A simple time delay system becomes infinite-dimensional and may produce a variety of rich complex dynamical behavior [1–9]. Therefore it is very important to explore the collective dynamical features of nonlinear time-delayed oscillators. To this end here we investigate the stability property of an injection synchronized OEO. Neyer and Voges introduced OEO for the first time [10]. Subsequently, Yao and Maleki proposed OEO as a high Q oscillator capable of producing pure microwave signal [11]. For the last few decades, an intensified effort has been imposed on either developing the architecture of OEO or exploring the dynamics of OEO as an intrinsic time-delayed oscillator, to cite a few Dual Loop Optoelectronic oscillator [12, 13], injection-locked dual OEO [14]. Different studies demonstrated the OEO without electrical BPF [15, 16]. A new configuration of OEO was presented by the present author where using the injection locking technique the electrical BPF is replaced by a van der Pol oscillator [17, 18]. Several investigations have been conducted to explore the collective dynamical behavior of the oscillator. Emergence of breathers in OEO was demonstrated by Chembo et al. in this study feedback strength was considered as the control parameter [19]. Several techniques have been reported to explore the generation of chaos and stability analysis in an OEO [20–25]. Considering both feedback loop gain and delay as control parameter the nonlinear dynamical behavior of an OEO was reported [26, 27] however, to design the delay and the electrical filter DSP technology was implimented.

In the present work we inject an external periodic signal to an OEO and consider the injection signal amplitude as a control parameter. The effect of the external periodic signal on the stability of the OEO is demonstrated numerically and analytically. Although suppression of chaos using sync signal is not new [30–33], however, to the best of our knowledge the effect of the external signal on the stability of the OEO is not addressed before. Here the OEO under test is similar to the basic OEO, proposed by Yao and Maleki. The application of sync signal has huge number of benefits; such as in OEO a long fiber delay creates unwanted cavity modes, these modes can be forbidden to emerge in the bandwidth of the filter with the application of injection signal. Therefore long optical delay lines can be used which results in reduction of the phase noise. Moreover, application of sync signal effectively increases the Q factor of the oscillator. In recent days synchronized OEO finds important application in IoT, such as OEO finds important application for high speed signal processing in IoT devices, and it may function as an RF source with reduced phase noise in IoT devices, OEO can be used for precise measurement and sensing in IoT devices, for navigation and object detection in IoT, etc. Therefore proper understanding of stability boundaries of a synchronized OEO is important to make it instrumental in various applications.

The organization of the chapter is as follows: The delay dynamic model equation and the basic configuration are described in Sect. 2 describes. In Sect. 3 the stability and bifurcation analysis of the driven OEO is reported. In Sect. 4 brief discussion on the application of injection-locked OEO (ILOEO) is presented. Finally the paper concludes in Sect. 5.

2 The Model

The schematic of an SLOEO is presented in Figs. 1 and 2 demonstrates its equivalent model. The SLOEO contains a Mach–Zehnder modulator (MZM) which modulates the optical output of the CW laser with its nonlinear transfer function. The intensity-modulated optical signal then moves along a long optical fiber delay-line. A photodetector subsequently converts the optical signal into an RF signal which is then amplified and filtered by a narrow band RF BPF about the microwave signal to be produced, and connected to the electrode of the MZM. In this section we derive the model equation of the SLOEO.

The relationship between output power of the MZM and the oscillator input voltage is [29]

$$P_M(t) = \frac{1}{2}\sigma P_0\left[1 - \eta_e \sin\pi\left(\frac{V_i(t) + V_B}{V_\pi}\right)\right] \tag{1}$$

where $V_i(t)$ is any existing stationary RF signal expressed as $V_i(t) = V(t)e^{j(\omega_0 t+\theta(t))}$, here P_0 is optical power, σ denotes the insertion loss of the MZM, the extinction

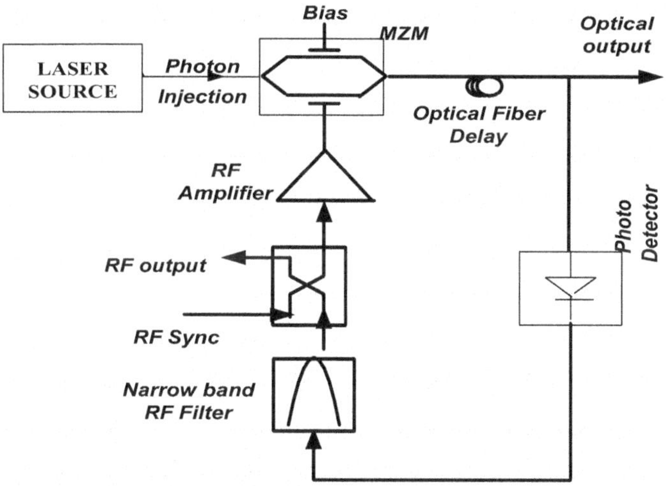

Fig. 1 Schematic representation of single loop optoelectronic oscillator (SLOEO)

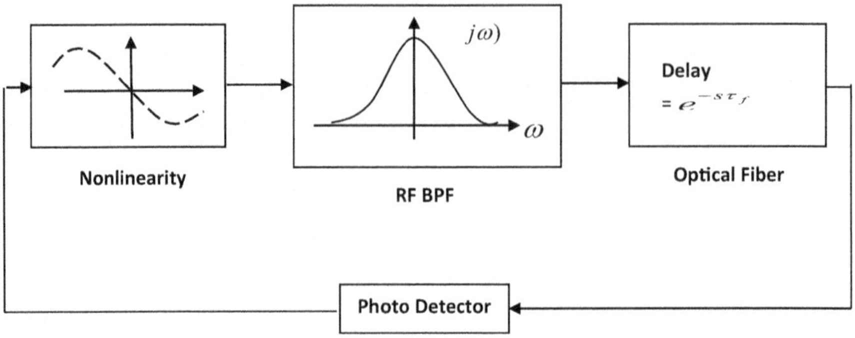

Fig. 2 Equivalent circuit model of SLOEO

ratio of the MZM is η_e, V_B is the bias voltage and V_π represents the half wave voltage of the MZM.

The RF output voltage is $V_0(t) = \rho_p R_p P_M(t - \tau_f)$, here the sensitivity of the photodetector is denoted by ρ_p, the photodetector impedance is represented by R_p and the feedback delay due to the optical fiber is represented by τ_f. Therefore the output of the photodetector can be expressed as [28, 29]

$$V_0(t) = V_{ph}\left[\begin{array}{l} 1 - \eta_e \sin\left(\frac{\pi V_B}{V_\pi}\right)\left\{J_0\left(\frac{\pi V(t-\tau_f)}{V_\pi}\right) + 2\sum_{m=1}^{\infty} J_{2m}\left(\frac{\pi V(t-\tau_f)}{V_\pi}\right)\cos\left[2m\omega\left(t-\tau_f\right)\right]\right\} \\ -2\eta_e \cos\left(\frac{\pi V_B}{V_\pi}\right) \times 2\sum_{m=0}^{\infty} J_{2m+1}\left(\frac{\pi V(t-\tau_f)}{V_\pi}\right)\sin\left[(2m+1)\omega\left(t-\tau_f\right)\right] \end{array}\right]$$

It is important to mention here that the pass band of the tuned circuit becomes narrow as the oscillation amplitude grows and the smaller frequency components are filtered out. The highest frequency term only sustains at the oscillator output. Therefore the MZM output can be expressed as

$$\begin{aligned} V_0(t) &= -2\eta_e V_{ph}\cos\left(\frac{\pi V_B}{V_\pi}\right) J_1\left(\frac{\pi V(t-\tau_f)}{V_\pi}\right)\sin\left[\omega\left(t-\tau_f\right)\right] \\ &= \frac{N\left(V\left(t-\tau_f\right)\right)}{V(t)}\exp\left(-s\tau_f\right) V_{in}(t) \end{aligned}$$

where $N[V(t-\tau_f)] = -2\eta V_{ph}\cos\left(\frac{\pi V_B}{V_\pi}\right)J_1\left(\frac{\pi V(t-\tau_f)}{V_\pi}\right)$ and $V_{ph} = \frac{\alpha R_P \rho_P P_0}{2}$, now for straightforwardness, we consider $\eta_e = 1$; $V_B = V_\pi$; $\pi V_{ph} = V_\pi$; $V_\pi = \pi$ and $N[V(t-\tau_f)] = 2J_1[V(t-\tau_f)]$, J_1 is the Bessel function of the first kind of order one. The RF oscillation voltage at the output of the oscillator can be expressed as

$$V_0(t) = \beta(s)\, V_{in}(t) \tag{2}$$

$$\beta(s) = \left[\frac{N[V(t-\tau_f)]}{V}G(s)e^{-s\tau_f}\right] \tag{3}$$

The quantity $G(s)$ corresponds to the transfer function of the frequency selective circuit and expressed as $G(s) = g_m Z(s)$, g_m denotes the gain of the frequency selective circuit. Using (2) and (3) we obtain

$$\frac{1}{Z(s)} = 2g_m\left[\frac{J_1[V(t-\tau_f)]}{V}e^{-s\tau_f}\right] \tag{4}$$

$$\frac{V}{Z(s)} \cong \left[2J_1[V(t-\tau_f)]\{\cos(\omega_0\tau_f) + j\sin(\omega_0\tau_f)\}\right]g_m \tag{5}$$

$$\left[\frac{V}{R} + C\frac{dV}{dt} + \frac{1}{L}\int Vdt\right] = [2J_1[V(t-\tau_f)]\{\cos(\omega_0\tau_f) + j\sin(\omega_0\tau_f)\}]g_m \tag{6}$$

$$\left(\frac{1}{R} + j\omega C + \frac{1}{j\omega L}\right) \cong G + 2C(j\omega - j\omega_0) \tag{7}$$

and

$$j\omega = \frac{1}{V(t)}\cdot\frac{dV}{dt} + j\omega_0 + j\frac{d\theta}{dt} \tag{8}$$

$$\frac{1}{R} + \frac{1}{V(t)}\frac{dV}{dt}\frac{2Q}{\omega_0 R} = g_m\frac{2J_1[V(t-\tau_f)]\cos(\omega_0\tau_f)}{V} \tag{9}$$

using (7) to (9) the amplitude equation of SLOEO bocomes

$$\frac{dV}{dt} = \frac{\omega_0}{2Q}[G_1 2J_1\left[V\left(t-\tau_f\right)\right]\cos(\omega_0\tau_f) - V(t)] \tag{10}$$

where $G_1 = g_m R$ is gain at resonance. Using (10) the building up of oscillation in OEO is depicted in Figs. 3 and 4 for the different values of Gain. Figure 5 shows the corresponding RF oscillation spectrum. Now if the delay τ_f is such that, $\omega_0\tau_f = 2n\pi$, $e^{-s\tau_f} = e^{-j\omega_0\tau_f} = 1$, it is not difficult to show that (9) can be written in the following form

$$\frac{dV}{dt} = \frac{\omega_0}{2Q}[G_1 2J_1\left[V\left(t-\tau_f\right)\right] - V(t)] \tag{11}$$

Now using the following parameters (11) can be rewritten as

$$t_n = \frac{\omega_0 t}{2Q},\quad \tau_n = \frac{\tau_f}{RC},\quad b = 2G_1,\quad v = \frac{V(t)}{V_{\max}},\quad v(t_n - \tau_n) = \frac{V(t-\tau_f)}{V_{\max}}$$

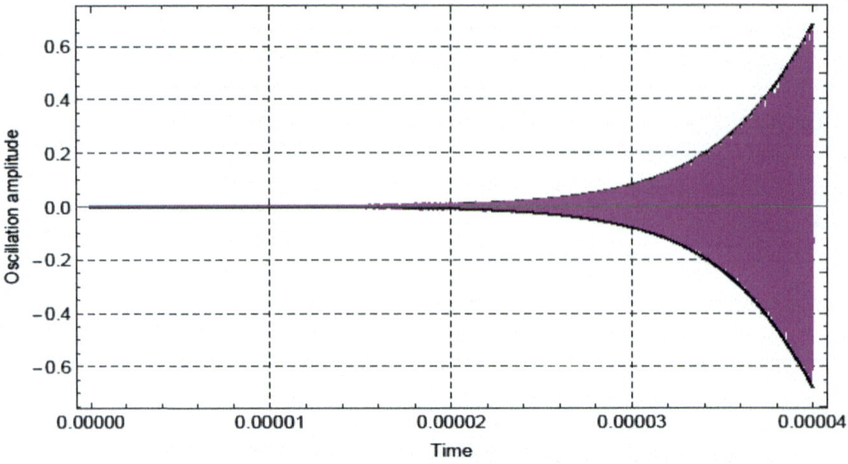

Fig. 3 Building up of RF oscillation at the output of the OEO. The parameter values are: $G_1 = 1.5$, $\tau_f = 1\,ns$, Q = 77

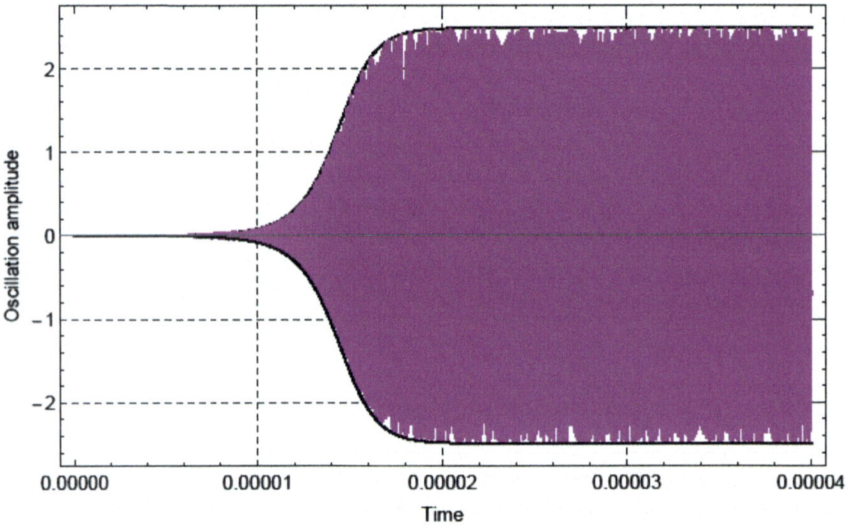

Fig. 4 Building up of RF oscillation at the output of the OEO. The parameter values are: $G_1 = 2.5$, $\tau_f = 1\,ns$, Q = 77

$$\frac{dv}{dt_n} = -v + bJ_1[v(t_n - \tau_n)] \tag{12}$$

Equation (12) is the normalized governing equation of SLOEO. Under the steady state condition ($\frac{dV}{dt} = 0$) and (11) will take the following form

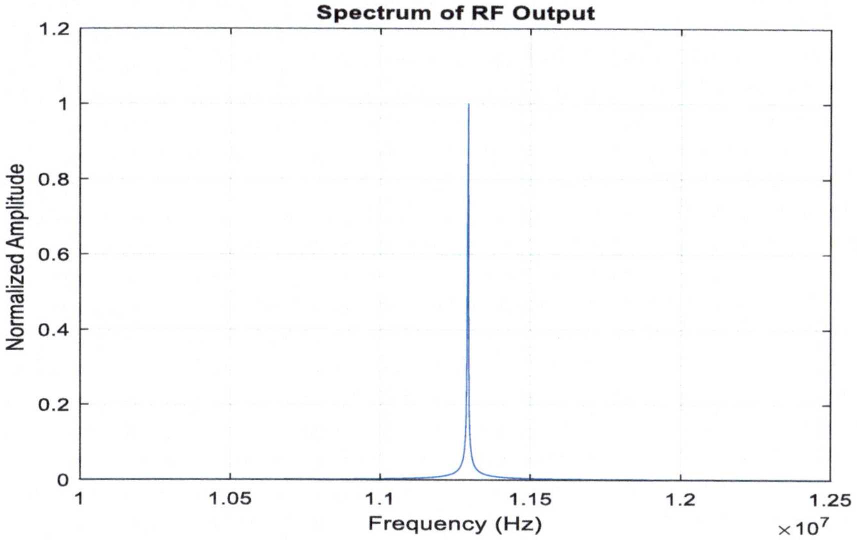

Fig. 5 RF output spectrum of SLOEO

$$\frac{\omega_0}{2Q}[G_1 2J_1[V(t_n)] - V(t_n)] = 0 \tag{13}$$

Decomposing $J_1[V(t)]$ in series and neglecting the higher order terms it is easy to observe that the steady-state amplitude of the oscillator can be expressed in the following form

$$V_{OSC} = 2\sqrt{2}\sqrt{1 - \frac{1}{G}} \tag{14}$$

3 Stability Analysis of an Injection Locked OEO (ILOEO)

In this section we employ an external periodic signal of the form $S(t) = Ee^{j(\omega_1 t + \psi(t))}$ to the undriven OEO, here E and $\psi(t)$ denote the amplitude and phase of the external signal respectively. $\phi(t) = \psi(t) - \theta(t)$ is the instantaneous phase difference between the free running and the external signal. Thus one can easily have the coupled voltage and phase governing equations of the driven OEO in the following form

$$\begin{aligned} \frac{dv}{dt_n} &= -v + bJ_1[v(t_n - \tau_n)] + G_1\, e \cos(\phi) \\ \frac{d\phi}{dt_n} &= \frac{2Q}{\omega_0}\Omega - \frac{G_1\, e}{v} \sin(\phi) \end{aligned} \tag{15}$$

The quantity e is the normalized strength of the external signal and $\Omega = \omega_1 - \omega_0$. Figure 6 demonstrates the lock range variation with injection signal strength. The figure clearly demonstrates that the lock range increases with the increment of the sync signal amplitude.

In order to explore the stability, the expression for the deviations (m, n) from the steady state (v, ϕ) are given by

$$\det \begin{pmatrix} m \\ n \end{pmatrix} = \begin{pmatrix} -\frac{3\,v^2}{8} & -G\,e\sin(\phi) \\ -v^2 & -\frac{G\,e}{v}\cos(\phi) \end{pmatrix} \begin{pmatrix} m \\ n \end{pmatrix} \tag{16}$$

We consider

$$a_{11} = -\frac{3\,v^2}{8} \tag{17}$$

$$a_{12} = -G\;e\;\sin(\phi) \tag{18}$$

$$a_{21} = -v^2 \tag{19}$$

$$a_{22} = -\frac{G\,e}{v}\cos(\phi) \tag{20}$$

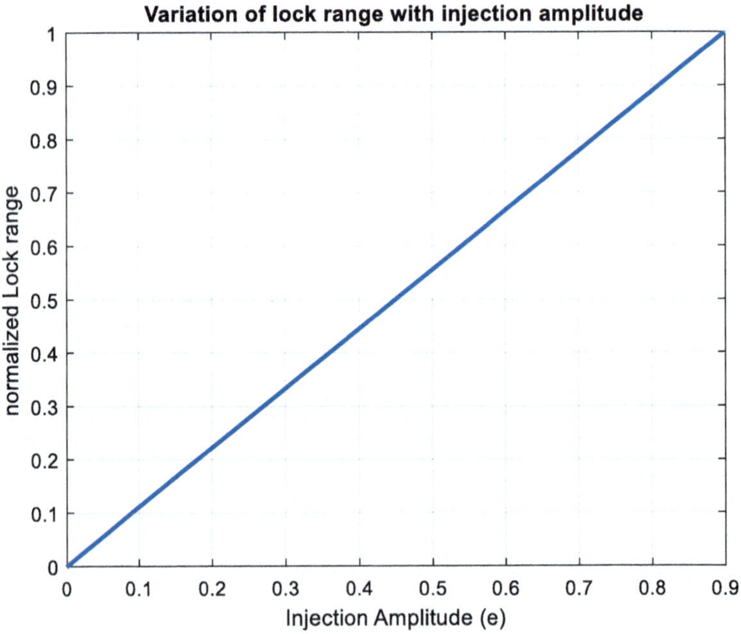

Fig. 6 Synchronization range and sync signal amplitude

The characteristic equation becomes

$$\begin{vmatrix} \lambda - a_{11} & a_{12} \\ a_{21} & \lambda + a_{22} \end{vmatrix} = 0 \tag{21}$$

In quadratic form

$$\lambda^2 + y\lambda + z = 0 \tag{22}$$

$$y = a_{11} - a_{22} \tag{23}$$

$$z = a_{11}a_{22} - a_{12}a_{21} \tag{24}$$

For Hopf bifurcation the necessary condition is

$$z = 0 \tag{25}$$

The influence of an injecting signal on the stability of an OEO is investigated. We solve (15) numerically and demonstrate the one parameter bifurcation diagram in Figs. 7 and 8. The external signal frequency is similar to the centre frequency of the undriven oscillator, i.e. we observe the dynamics under the locked condition $\Omega = 0$. The normalized injection strength e is varied as a control parameter. It is apparent from Fig. 7 that for $e = 0.65$ we get period doubling bifurcation and for $e = 1.2$ and $e = 1.4$ we get hopf bifurcation, here we consider the normalized feedback delay $\tau_n = 3$ and keep all other parameter values fixed. Further in Fig. 8 we change the normalized feedback delay $\tau_n = 3$ and obtain a more complicated dynamics.

4 Applications of ILOEO

The ILOEO is a fascinating technology with potential applications in the Internet of Things (IoT) realm. Here are some potential applications:

a. Wireless Sensor Networks (WSNs): WSNs in IoT often require low-power, low-cost, and robust communication systems. ILOEO can serve as a stable, high-frequency signal source for wireless sensor nodes, enabling synchronized operation and efficient communication among them.
b. Timing and Synchronization: In IoT applications where precise timing and synchronization are crucial, such as industrial automation, smart grids, and environmental monitoring, ILOEO can provide highly stable and synchronized clock signals. This ensures accurate data collection, coordination among devices, and efficient energy management.

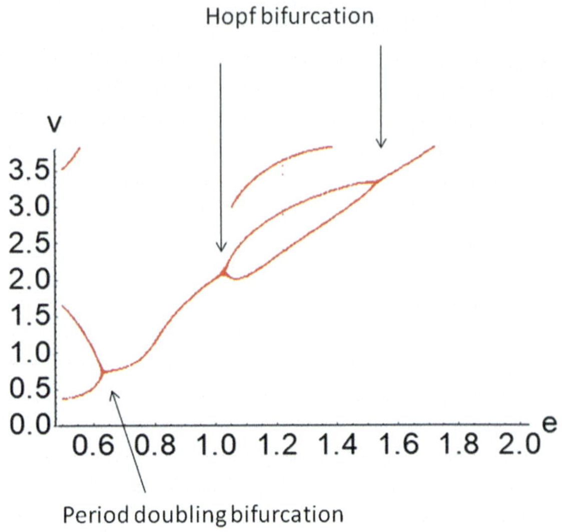

Fig. 7 Bifurcation diagram of the synchronized OEO here e is the control parameter: $\tau_n = 2$, all otherparameter values are fixed

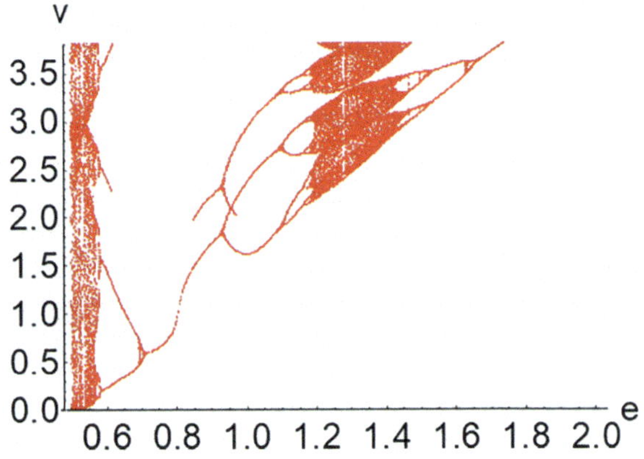

Fig. 8 Bifurcation diagram of the synchronized OEO here e is normalized injection strength: $\tau_n = 3$, all other parameter values are fixed

c. Radio Frequency Identification (RFID): RFID systems play a vital role in IoT for tracking and identifying objects. ILOEO can enhance the performance of RFID systems by providing a stable clock signal for reader devices, improving read range, data accuracy, and reliability.
d. Smart Agriculture: In precision agriculture, IoT devices are used for monitoring soil moisture, temperature, and other environmental parameters. ILOEO can be employed in sensor nodes to provide precise timing for data collection and transmission, ensuring optimal resource management and crop yield.
e. Smart Home Automation: In smart homes, various IoT devices like sensors, actuators, and smart appliances need to communicate and operate seamlessly. ILOEO can contribute to the synchronization of these devices, enabling efficient control, energy savings, and enhanced user experience.

f. Health Monitoring and Wearable Devices: Wearable IoT devices for health monitoring require accurate timing for data sampling and transmission. ILOEO can provide stable clock signals for synchronization among wearable sensors, ensuring real-time monitoring of vital signs and timely health alerts.
g. Industrial IoT (IIoT): In industrial settings, IIoT applications demand precise timing and synchronization for processes like distributed control, automation, and monitoring. ILOEO can serve as a reliable clock source for industrial IoT devices, improving system efficiency, productivity, and safety.
h. Environmental Monitoring: IoT devices are extensively used for environmental monitoring in smart cities, pollution control, and natural disaster management. ILOEO can facilitate synchronized operation among distributed sensor nodes, enabling accurate data collection and analysis for timely decision-making.

5 Conclusions

In this work we propose a delay differential equation of the injection synchronized OEO and investigate the stability properties analytically and numerically. We show for the first time that interplays between feedback loop delay, nonlinearity, and injected signal strength leads to Hopf bifurcation and period doubling bifurcation in the system. We discuss the bifurcation diagrams in terms of the Injection strength and feedback loop delay. Initially we derive the system equation of the OEO and observe the growth of oscillation with variation of the feedback gain. We then employ the external periodic sync signal to the free running OEO. In order to explore the stability analysis we vary the injected signal amplitude as a control parameter. It has been observed through detailed numerical and analytical study that with variation of the sync signal, and for different values of the feedback loop delay, Hopf bifurcation and period doubling bifurcation and even more complicated dynamics are produced by the oscillator. The ability of producing extremely pure microwave signals makes OEO instrumental in many applications such as deep space mission, 5G communication, as a stable local oscillator in RADAR, high performance analog to digital converters, in GPS. Nowadays integration of OEO is also done with IoT for air cargo security, which offers significant potential for enhancing the efficiency and effectiveness of cargo inspection and monitoring processes. Injection-locked oscillators are widely used to reduce the phase noise, injection locking improves the long-term stability of the OEO. OEO can be used as a voltage controlled oscillator (VCO) with both optical and electrical output. Therefore identifying stability boundary of a synchronized OEO is important to make it functional in various applications. Overall, the Injection Locked Optoelectronic Oscillator holds promise for various IoT applications by providing stable timing signals, enhancing communication reliability, and improving the overall performance and efficiency of IoT systems.

Acknowledgements The authors are thankful to the management of Siliguri Institute of Technology, Siliguri, West Bengal, for carrying out the work.

References

1. Ikeda K, Mutsumoto K (1987) High dimensional chaotic behavior in systems with time delayed feedback. Phys D 29:223
2. Ikeda K, Kondo K (1982) Successive higher-harmonic bifurcation in systems with delayed feedback. Phys Rev Lett 49:1467
3. Sprott JC (2007) A simple chaotic delay differential equation. Phys Lett A 366
4. Mackey MC, Glass L (1977) Oscillation and chaos in physiological control system. Science 197:287–289
5. Lakshmanan M, Senthilkumar DV (2011) Dynamics of nonlinear time delay systems, 1st edn. Springer
6. Senthilkumar DV, Lakshmanan M (2005) Bifurcations and chaos in time delayed piecewise linear dynamical systems. Int J Bifurc Chaos 15:2895
7. Hu G, Jiang S (2008) Generating hyperchaotic attractors via approximate time delayedstate feedback. Int J Bifurc Chaos 18:3485–3494
8. Perez G, Cerdeira H (1995) Extracting messages masked by chaos. Phys Rev Lett 74:1970
9. Goedgebuer JP, Larger L, Porte H (1998) Optical cryptosystem based on synchronization of hyper-chaos generated by a delayed feedback tunable laser diode. Phys Rev Lett 80:2249–2252
10. Neyer A, Voges E (1982) High-frequency electro-optic oscillator using an integrated interferometer. Appl Phys Lett 40:6–8
11. Yao XS, Maleki L (1996) Optoelectronic microwave oscillator. J Opt Soc Am B 13:1725–1735
12. Yao XS, Maleki L (1997) Dual microwave and optical oscillator. Opt Lett 22:1867–1869
13. Yao S, Maleki L (2000) Multi-loop optoelectronic oscillator. IEEE J Quantum Electron 36:79–84
14. Zhou W, Blasche G (2005) Injection locked dual opto-electronic oscillator with ultra-low phase noise and ultra-low spurious level. IEEE Trans Microw Theory Tech 53:13296–13301
15. Yao XS (1997) High-quality microwave signal generation by use of Brillouin scatter-ing in optical fibers. Opt Lett 22:1329–1331
16. Liu X, Pan W, Zou X, Luo B, Yan L, Lu B (2012) A reconfigurable optoelectronic oscillator based on cascaded coherence-controllable recirculating delay lines. Opt Express 20:13296–13301
17. Mukherjee A, Ghosh D, Das NR, Biswas BN (2016) Harmonic distortion and power relations in a single loop opto-electronic oscillator with RF band pass filter replaced by an injection locked oscillator. Opt Int J Light Electron Opt 127:973
18. Ghosh D, Mukherjee A, Chatterjee S, Biswas BN (2016) A comprehensive theoretical study of dual loop opto electronic oscillator. Opt Int J Light Electron Opt Elsevier 127:3337
19. Chembo Koumou Y, Colet P, Larger L, Gastaud N (2005) Chaotic breathers in delayed electro-optical systems. Phys Rev Lett 95:203903
20. Chembo YK, Larger L, Colet P (2008) Nonlinear dynamics and spectral stability of optoelectronic microwave oscillator. IEEE J Quantum Electron 44:858–868
21. Chembo YK, Larger L, Tavernier H, Bendoula R, Rubiola E, Colet P (2007) Dynamic stabilities of microwaves generated with optoelectronic oscillators. Opt Lett 32:2571–2573
22. Peil M, Jacquot M, Kouomou YC, Larger L, Erneux T (2009) Routes to chaos and multiple time scale dynamics in broadband bandpass nonlinear delay electro-optic oscillators. Phys Rev E 79:026208
23. Weicker L, Erneux T, Jacquot M, Chembo Y, Larger L (2012) Crenelated fast oscillatory outputs of a two-delay electro-optic oscillator. Phys Rev E 85:026206
24. Larger L, Lacourt P-A, Poinsot S, Hanna M (2005) From flow to map in an experimental high dimensional electro-optic nonlinear delay oscillator. Phys Rev Lett 95:043903
25. Callan KE, Illing L, Gauthier DJ, Scholl E (2010) Broad band chaos generated by an optoelectronic oscillator. Phys Rev Lett 104
26. Murphy TE et al (2010) Complex dynamics and synchronization of delayed feedback nonlinear oscillators. Proc R Soc Lond A 368:343–366

27. Cohen AB, Ravoori B, Murphy TE, Roy R (2008) Using synchronization for prediction of high dimensional chaotic dynamics. Phys Rev Lett 101:154102
28. Biswas BN, Chatterjee S, Pal S (2012) Laser induced microwave oscillator. IJECET 3:211–219
29. Mukherjee A, Biswas BN, Das NR (2015) A study on the effect of synchronization by an angle modulated signal in a single loop optoelectronic oscillator. Opt Int J Light Electron Opt Elsevier 126:1815
30. Pyragas K (1992) Continuous control of chaosby self controlling feedback. Phys Lett A
31. Piserchik AN, Goswami BK (2000) Annihilation of one of the coexisting attractors in a bistable system. Phys Rev Lett
32. Martinez-Zerega BE, Piserchik AN (2005) Efficiency of the control of coexisting by harmonic modulation applied in different ways. Phys Lett
33. Bleich ME, Socolar JES (1995) Stability of periodic orbits controlled by time-delay feedback. Phys Lett

Chapter 10
An EMG-Based Cost-Effective Prosthetic Hand

Rajdeep Saha, Abhijit Biswas, Saradwat Sen, and Sayani Dhali

Abstract In the era of assistive technology, the need for affordable and accessible prosthetic solutions has been roaming around forever. We introduce the solution by making an inexpensive prosthetic arm controlled by electromyographic (EMG) signals obtained from the upper arm muscles of amputees. This project addresses a major global issue which is, in an estimate over 100 million individuals worldwide require prosthetic limbs due to limb damage or amputation. Unfortunately, a large portion of these individuals (especially those from lower economic backgrounds) face problems accessing due to the high cost of traditional prosthetic options, which can range from Rs. 70,000 to Rs. 3.5 lakh for imported prosthetic hands. But this project is Remarkably cost-effective. The overall system costs around a few thousand rupees significantly reducing expenses compared to market alternatives. This project solves the solution by using cost-effective technologies in the most efficient ways. The key component in this prosthetic hand is the Muscle EMG sensor. The noise reduction and the amplification of the Bio-Potential signal are done in the IC of the EMG sensor, which helps in precise signal acquisition. By integrating this sensor into the prosthetic hand design, the component can accurately capture muscle signals and translate them into commands for prosthetic hand movements. The brain of this prosthetic hand is a microcontroller, which serves as the central processing unit for signal processing as Band-Pass Butterworth IIR digital filter and responsiveness. Through confined algorithms and real-time data analysis, the microcontroller reads EMG signals captured by the sensor, enabling full control of the prosthetic hand. This integration of hardware and software makes the project's goal of affordability without compromising performance possible. Beyond the technical aspects, this project solves the topmost prioritized problem which is affordability with efficiency. In third world countries where a large portion of people are suffering through poverty, surgery on amputation is nowhere near possible, let alone installing prosthetics. This cheap but effective alternative which costs around 1–3% of the market's product cost, can be affordable to the mass population with disabilities and enhance their quality of life.

R. Saha (✉) · A. Biswas · S. Sen · S. Dhali
Department of Electronics and Communications Engineering, Supreme Knowledge Foundation Group of Institution, 1, Khan Road, Mankundu, Hooghly, West Bengal 712139, India
e-mail: rajdeeps.9123080971@gmail.com

A. Acharyya et al. (eds.), *Real-World Applications and Implementations of IoT*, Studies in Smart Technologies, https://doi.org/10.1007/978-981-97-8627-5_10

Keywords Electromyographic signals · Bio-potential amplifier · Band-pass butterworth IIR digital filter · Cost-effective technology · Microcontroller integration · Affordable prosthetic solutions

1 Introduction

Prosthetics are artificial devices that compensate for the lack of existence of limbs throughout the body. A prosthetic arm replaces the amputated or inoperative upper limbs. This offers individuals who have lost limbs a chance to regain mobility and independence. Now unlike traditional prosthetic arms, a Myoelectric Prosthetic arm is controllable by myoelectric (EMG) signals expediently. This is made achievable by the fact that amputees' neuro-muscular systems continue to function normally even after amputation. After appropriate processing, the leftover signals are utilized and are sufficient to control the arm's movement [1]. The technical approach of Sudesh Garg, Kiran Rathi, and Amit Kr. Bansal is concentrated on creating a 3D-printed bionic arm that prioritizes human control via EMG sensors, affordability, and usefulness. It is possible to fabricate objects that are lightweight and customizable by using PLA and 3D printing technology. Evaluating for necessary hand movements guarantees usability and imitation of natural hand movements. The objective is to offer amputees an affordable and easily obtainable option that prioritizes controllability and enhanced quality of life. However, noise filtering methods are not specifically mentioned [2]. Furthermore, to separate muscle signals from outside interference in EMG-based systems, noise filtering techniques are crucial for precise bionic arm control as well as improved user experience and safety.

The research aim focuses on decreasing external noise interference that occurs during signal transmission from the brain to the skin of the targeted muscle location. We successfully reduced this interference by utilizing the Butterworth digital filter, which guarantees more precise and more accurate signal capture. This advanced filtering system is indispensable to improve the prosthetic device's overall reliability and effectiveness.

The merit of the following research lies within the cost factor of the prosthetic arm. It has been estimated that there are roughly 0.62 amputees in India per thousand population [3, 4]. With that number of amputees, we are aware that the traditional EMG-controlled prosthetic's pricings are way beyond affordable for the lower-income groups. This cheap but efficient controllable prosthetic arm will be viable for every individual regardless of their monetary situation.

In addition, we have added a manually operated radical feature that makes it quite effortless for users to hold a pen and write. With the help of this unique feature, people with upper limb disabilities can now write with greater confidence and ease.

2 Methodology

The initial step involves using dry surface electrodes to gather EMG data from the patient's hand. For EMG sensors, three electrodes are employed: two for recording the signal and one as a reference [5]. When targeting specific muscles like the forearm, one electrode is placed at the muscle's center, and the other at a distance of 1.5 inches. The reference electrode is positioned on the bony side or the back side of the forearm.

To minimize electrode-to-skin impedance, the electrodes are pre-gelled. Before applying the electrodes, the skin is sanitized [5]. The collected EMG data are then analyzed to detect forearm movement and distinguish between hand states such as flexion and extension [5, 6].

2.1 Acquisition of Signal Using Muscle BioAmp Band

Three electrodes are used here to capture the EMG signal from the muscle of the forearm [5]. Two of them are bipolar surface electrodes and one is a reference electrode [5]. Then the signal is sent to the Muscle BioAmp Candy through BioAmp Cable and then to the Arduino Nano for further processing (Fig. 1).

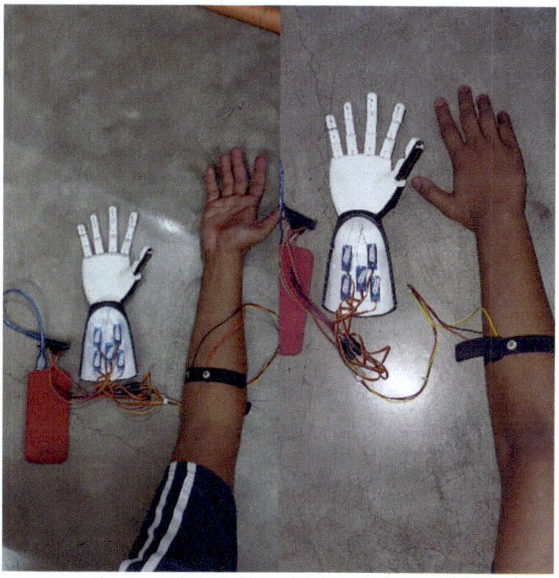

Fig. 1 Electrodes placement for signal acquisition [5]

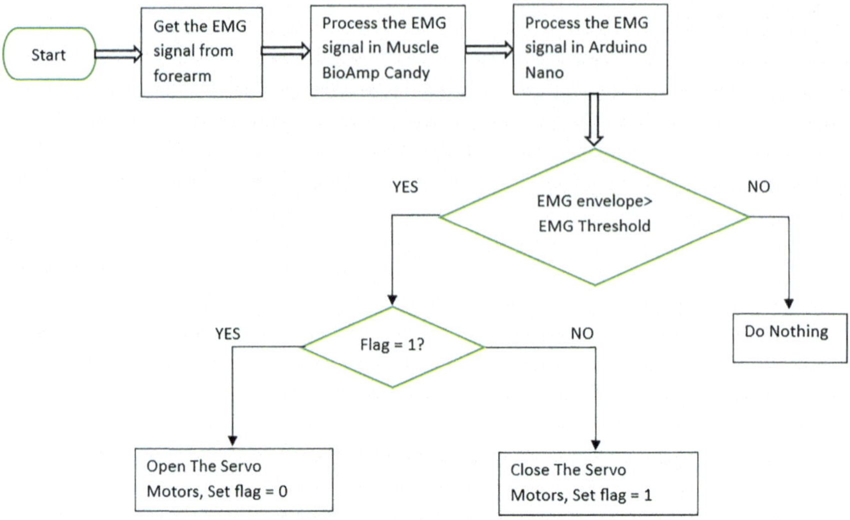

Fig. 2 Flow diagram of the process

2.2 *Control of the Prosthetic Hand*

The prosthetic hand was designed to assist the hand's elbow motions by using signals derived from wrist motions [5]. The EMG signals were processed and then delivered as control signals to an Arduino Nano. Wrist flexion is the signal that is obtained if the person closes their hand or flexes their forearm muscle.

This results in the motor rotating in an upward and downward direction, enabling the hands of individuals who are amputated to open and close the hand, and grasp objects. The Arduino Nano's PWM ports are connected to the motor's control input, which is powered by an external source.

Consequently, the prosthetic hand was successfully operated using EMG signals that were recorded from the healthy hand [6] (Fig. 2).

2.3 *Circuit Diagram*

Now, lets discuss about the circuitry and power management system utilized in our project. The whole system has 5 servo motors for accurate finger movement, an EMG sensor for muscle signal acquisition, and a power bank for portable operation. Each component's connections to the Arduino Nano microcontroller are discussed below in this project (Fig. 3).

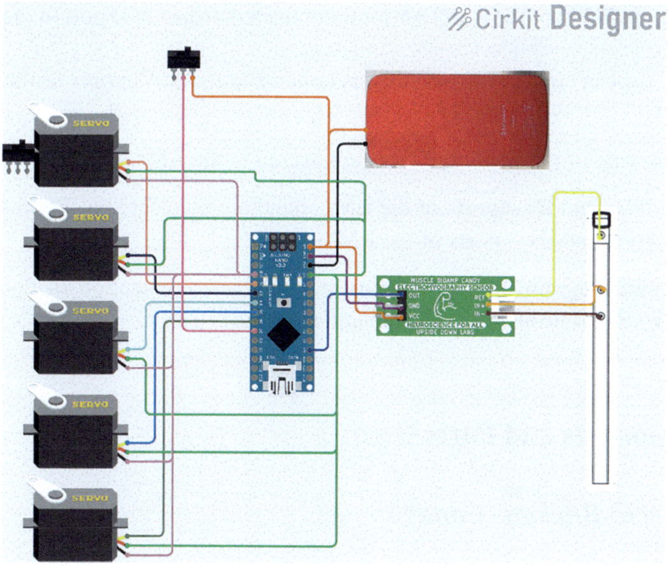

Fig. 3 Circuit diagram of the prosthetic hand

Servo Motors (Thumb, Index, Middle, Ring, Pinky):

- Each servo motor is connected to the Arduino Nano for control.
- Signal wires (usually PWM) from each servo are connected to digital pins (D2, D3, D4, D5, D6) of the Arduino Nano [7].
- Power for each servo is drawn from the 5 V port of the Arduino Nano.
- Ground connections for each servo are made to the GND port of the Arduino Nano.

Power Management (Power Bank):

- The power bank used is a "Reconnect Power Hub 10,000 mAh Power Bank Series 100".
- The power bank is connected to the Arduino Nano via its USB port, providing power to the entire system.

Write Switch:

- One toggle switch is connected from Vin to D7.
- By turning it on the hand will enable the user to hold a pen and write with it.

Muscle BioAmp Candy(EMG Sensor):

- The EMG sensor used is a "Muscle BioAmp Candy".
- The output side of the EMG sensor is connected to the Arduino Nano.
  - The analog output (Out) of the EMG sensor is connected to analog pin A0 of the Arduino Nano.

  - Ground (GND) of the EMG sensor is connected to the GND port of the Arduino Nano.
  - VCC (power) of the EMG sensor is connected to the Vin port of the Arduino Nano.
- The input side of the EMG sensor is connected to the EMG band.
  - REF, IN+, and IN− ports of the EMG band are connected accordingly to the REF, IN+, and IN− ports of the EMG sensor.

This circuit diagram effectively illustrates the interconnection of components within the EMG-controlled prosthetic hand system.

3 Components and Parts Used

3.1 Muscle-BioAmp-Candy

Muscle-BioAmp-Candy is a candy-sized Electro Myography (EMG) sensor designed to capture precise muscle bio-potential signals. With a compact size and functionality, this device can transform the way of monitoring muscle activity.

It has a fixed gain of x2420 and a Band Pass filter spanning from 72 to 720 Hz, Muscle-BioAmp-Candy ensures signal clarity and accuracy.

Compatible with standalone ADCs such as the ADS1115 or any microcontroller development board equipped with an ADC, Muscle-BioAmp-Candy offers seamless integration into your existing setup [8] (Figs. 4 and 5).

Fig. 4 Muscle-BioAmp-Candy [8]

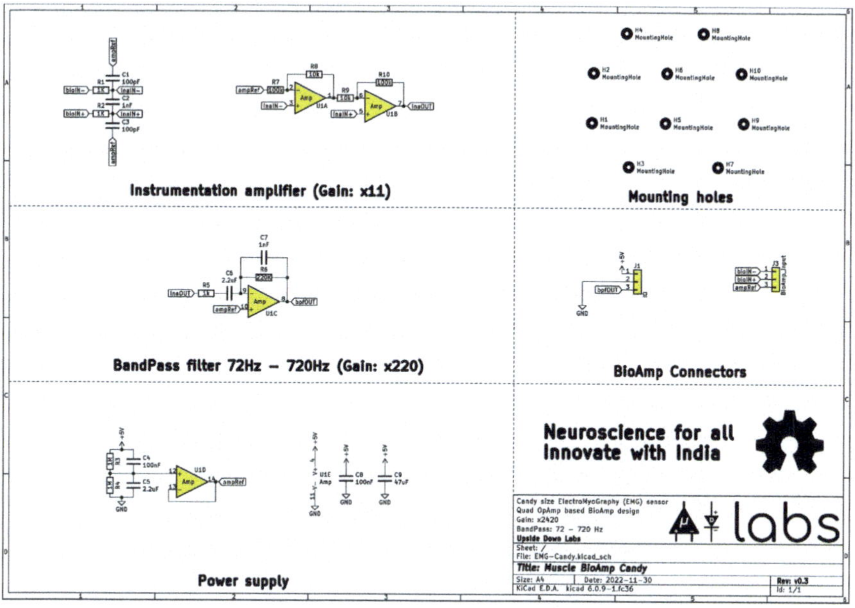

Fig. 5 Schematic diagram of Muscle-BioAmp-Candy [8]

3.2 *Schematic Diagram-*

3.3 *Muscle BioAmp Band (EMG Band)*

Stretchable in nature, the Muscle BioAmp Band (EMG Band) can be linked to the Muscle BioAmp-Candy via a BioAmp Cable that uses dry electrodes. It makes it simple for you to capture your muscle signals [9].

Dimensions

See Figs. 6 and 7.

3.4 *Arduino Nano*

Arduino Nano is a small, breadboard-friendly development board based on an ATmega 328P SMD package microcontroller and offers the same connectivity and specs as "Arduino Uno in a small package" [11]. In our project, we are using it because the size is the main factor. After all, the microcontroller needs to be fit into the 3d printed hand. We are supplying the power to the board with a Mini B USB port present on it [10, 11] (Fig. 8).

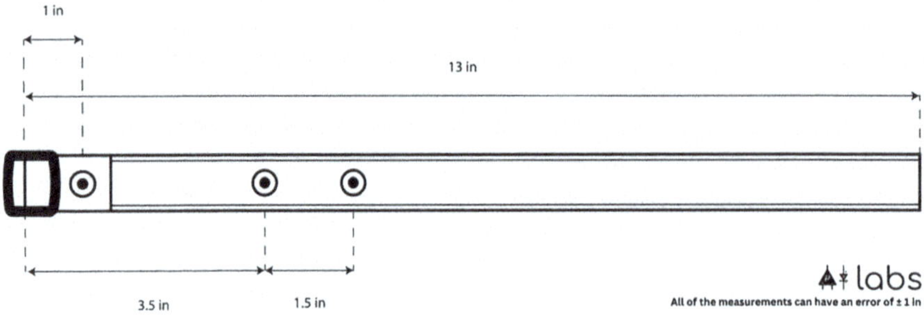

Fig. 6 Dimensions of the muscle BioAmp band (EMG Band) [9]

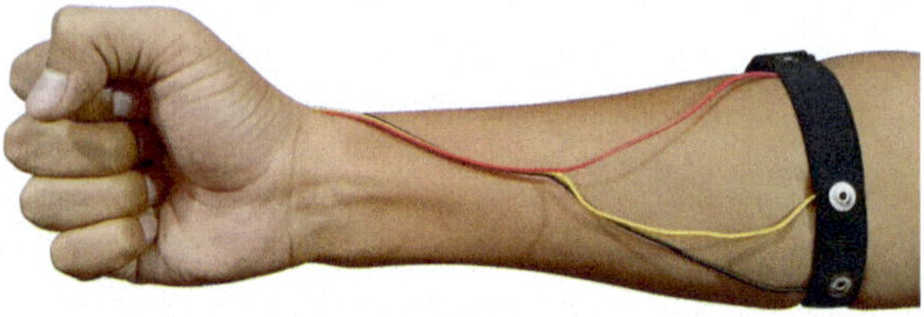

Fig. 7 Electrode placement example [9]

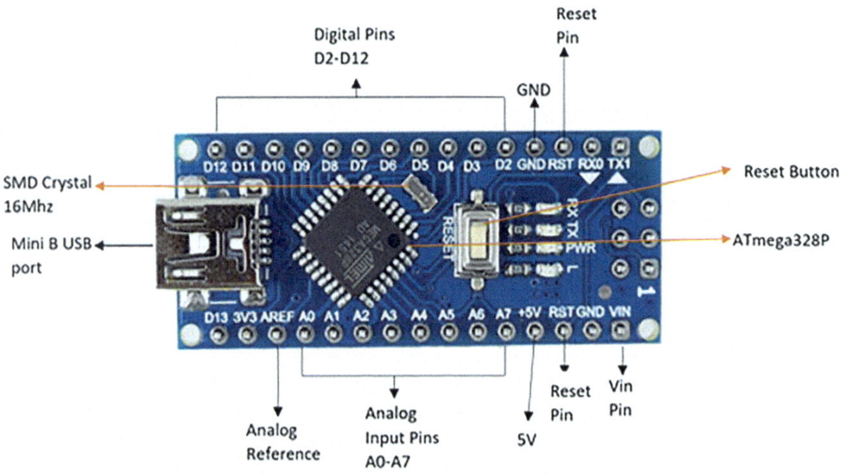

Fig. 8 Arduino nano board [10]

3.5 SG90 9 g Micro Servo [12]

SG90 9 g Micro Servo is tiny and lightweight with high output power [12]. The servo motors are used for the movement of the fingers in the hand. The servo motors usually provide control over the 180° range [11]. This angular position control is controlled by the PWM technique so by varying its duty cycle we can control the angular position of the motor [11]. This servo motor can lift a maximum of 1.6 kg when suspended at a 1 cm distance from the shaft [11, 13] (Figs. 9 and 10).

Every 20 ms, the servo examines the pulse [14]. The servo can be rotated to zero degrees by a pulse with a width of 1 ms (1 ms), to ninety degrees (the neutral position) by a pulse of 1.5 ms, and to 180° by a pulse of 2 ms [14, 15] (Fig. 11).

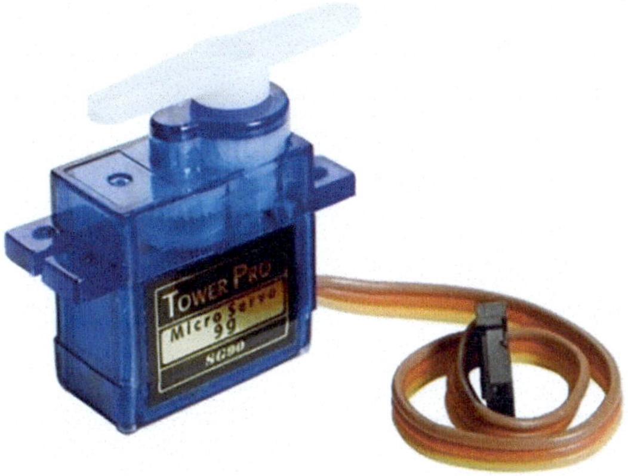

Fig. 9 SG90 9 g Micro Servo [12, 13]

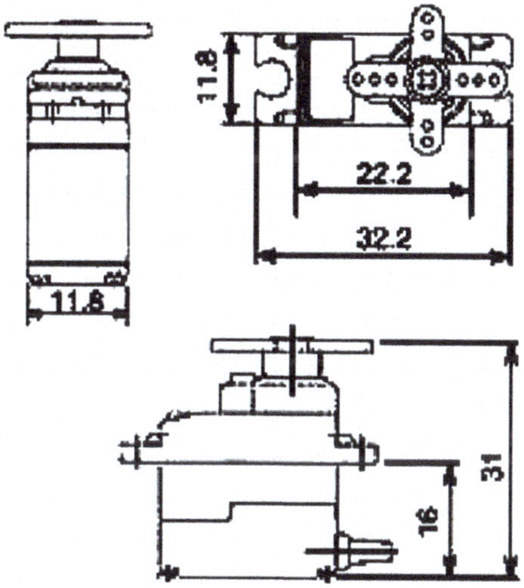

Fig. 10 Orthographic view of SG90 9 g Micro Servo [12, 13]

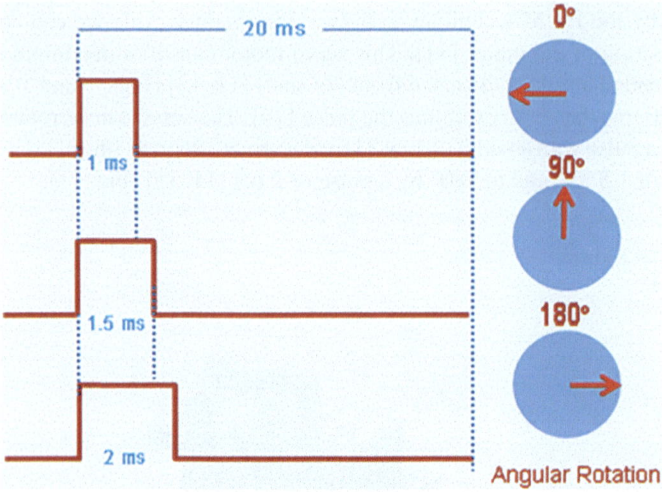

Fig. 11 Rotation control of SG90 9 g Micro Servo [15]

Fig. 12 Reconnect power hub 10,000 mAh power bank series 100 [16]

3.6 Power Bank

Here we are using Reconnect Power Hub 10,000 mAh Power Bank Series 100. It has a capacity of 10,000 mAh, so it can be used for longer intervals. As the microcontroller, servo, and Muscle-BioAmp-Candy draw very little power on a full charge it lasts for about 10 h [16] (Fig. 12).

3.7 Fabrication of 3D Printed Hand

Here we used PLA because PLA is a common 3D printing material to print components. PLA is widely used because of its ease of availability, ecological footprint, and affordability.

Table 1 Features of PLA [2, 17]

S. no	Property	Value
1	Melting point	Low (150 and 180 °C)
2	Thermal expansion	Low (68 μm/m K)
3	Layer adhesion	Moderate
4	Heat resistance	Low (55–65 °C)
5	Tensile strength [18]	High (39.9–52.5 MPa) [18]
6	Compressive strength [18]	High (48.2–62.0 MPa) [18]
7	Dimensional accuracy	High

Table 2 Features of 3D printer [19]

Model	Ender 3
Physical dimensions [20]	(w) 40 cm × (d) 22 cm × (h) 46 cm [20]
Maximum printing area [20]	(w) 20 cm × (d) 22 cm × (h) 22 cm [20]
Wire diameter	0.2 mm
Nozzle diameter	0.4 mm
Platform temperature [20]	~100 °C [20]
Nozzle printing temperature [20]	~200 °C [20]
Cooling method [20]	Air cool
Motor drive	Stepper motor drive

PLA is a biodegradable substance derived from food crops like sugarcane, corn, and jowar. Here are some key characteristics of PLA and the 3D printer [2] (Tables 1 and 2).

Now let us discuss the specification of the 3D Printed Hand (Figs. 13 and 14) (Table 3).

4 Algorithms and Their Detailed Analysis

4.1 Algorithm of the Arduino Code

Initialization:

- Set up the system parameters including the header files, sampling rate, baud rate, and pin configurations.
- Initialize the communication interfaces (e.g., Serial) for monitoring and debugging purposes.

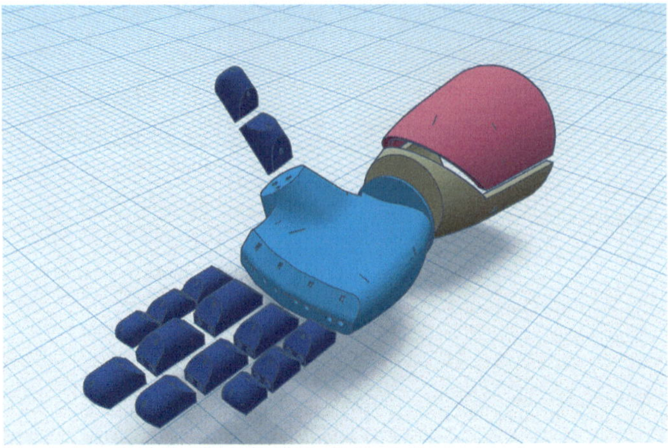

Fig. 13 Isometric projection of the 3D printed hand [19]

Fig. 14 After assembling all the parts of the hand

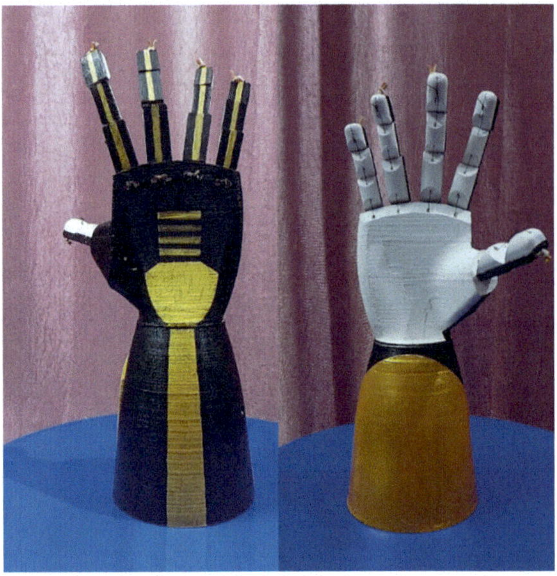

Setup:

- Attach the servo motors to their respective pins and set their initial positions.
- Define the WritePin.
- Configure the EMG sensor input pin and any auxiliary pins required for system operation.
- Define the threshold voltage (different for each person).

Table 3 Technical specification of 3D printed parts [21]

No	Name of the parts [22]	No. of joints/parts [22]	Length (cm) [22]
1	Thumb finger [22]	2 [22]	5.6 [22]
2	Index finger [22, 23]	3 [22, 23]	7.5
3	Middle finger [22, 23]	3 [22, 23]	8.2
4	Ring finger [22, 23]	3	6.8
5	Pinky finger	3	5.4
6	Palm	1	9
7	Wrist	–	11.5
8	The diameter of the end of the wrist	–	9.3
9	The total length of the arm	–	27.7

Main Loop:

- Check WritePin is High or Low.
  If High, Rotate the thumb, index, and middle finger as such it can hold a pen for writing purposes.
  If Low, Continue to the next steps.
- Continuously sample the EMG signal at the defined sampling rate.
- Apply a band-pass Butterworth IIR digital filter to the raw EMG signal to extract relevant muscle activity within the desired frequency range.
- Compute the EMG envelope using an envelope detection algorithm to estimate the magnitude of muscle activation.

Gesture Recognition:

- Compare the normalized envelope value with a predefined threshold to determine muscle activation and gesture recognition.
- Implement a hysteresis mechanism to prevent rapid toggling of the hand due to noise or minor fluctuations in muscle activity.
- Define specific thresholds for opening and closing gestures based on individual user characteristics and preferences.

Servo Control:

- Based on the detected gesture:
  - If the muscle activation exceeds the closing threshold:

    Close the hand by rotating the servo motors to the predefined closed position.
  - If the muscle activation falls below the opening threshold:

    Open the hand by rotating the servo motors to the predefined open position.
- Implement a gesture delay to prevent rapid and unintended toggling of the claw in response to minor fluctuations in muscle activity [8].

4.2 Algorithm for Envelope Detection

1. Subtract the previous EMG signal value from the sum.
2. Add the absolute value of the current EMG signal to the sum [24].
3. Store the absolute value of the current EMG signal in the circular buffer at the current index [25].
4. Update the data index to point to the next position in the circular buffer, wrapping around to the beginning if necessary.
5. Compute the average of the EMG signal values in the circular buffer by dividing the sum by the buffer size.
6. Multiply the average by 2 to scale the envelope signal.
7. Return the computed envelope signal.

4.3 Algorithm of EMG Band Pass Filter

Algorithm for the Band-Pass Butterworth IIR digital filter:

1. **Initialization**:
    - Initialize the state variables **z1** and **z2** for each filter section to zero.
2. **Filtering Process**:
    - For each input sample:
        - For each filter section:

            Calculate the intermediate variable **x** using the difference equation of a second-order IIR filter [26].
            Update the output using the calculated **x** value and the previous state variables.
            Update the state variables **z1** and **z2** for the next iteration.
3. **Output**:
    - Return the filtered output.

Here's a breakdown of the steps within the filtering process:

- For each filter section:
    1. Calculate the intermediate variable **x** using the difference equation:

    $$x = \text{input} - a1 * z1 \; - \; a2 * z2$$

    where **input** is the current input sample, **z1**, and **z2** are the previous state variables, and **a1** and **a2** are the filter coefficients.

2. Update the output using the calculated **x** value and the previous state variables:

$$\text{output} = \text{b0} * \text{x} + \text{b1} * \text{z1} + \text{b2} * \text{z2}$$

   where **b0**, **b1**, and **b2** are the filter coefficients for the output [27].
3. Update the state variables **z1** and **z2** for the next iteration:

$$\text{z2} = \text{z1 z1} = \text{x}$$

Repeating these steps for each input sample, obtained the filtered output of the Band-Pass Butterworth IIR digital filter [8, 28–30].

4.4 General Difference Equation of Band-Pass Butterworth IIR Digital Filter

The code implements a Band-Pass Butterworth IIR digital filter using second-order sections (biquads) [31]. Each biquad represents a second-order IIR filter section. Let's break down the mathematical expression for each biquad.

The general difference equation for a second-order IIR filter is [32]:

$$y[n] = b_0x[n] + b_1x[n-1] + b_2x[n-2] - a_1y[n-1] - a_2y[n-2]$$

where:

- $y[n]$ is the output at time n
- $x[n]$ is the input at time n
- $x[n-1]$ and $x[n-2]$ are the previous input samples
- $y[n-1]$ and $y[n-2]$ are the previous output samples
- b_0, b_1, b_2 are the feedforward (numerator) coefficients
- a_1, a_2 are the feedback (denominator) coefficients.

Let's express each biquad in the code as difference equations:

First Biquad:

$$\begin{aligned} \text{z1\_1[n]} &= \text{x[n]} - 0.05159732 * \text{z1\_1[n-1]} - 0.36347401 * \text{z2\_1[n-1]} \\ \text{z2\_1[n]} &= \text{z1\_1[n-1]}\,\text{y\_1[n]} = 0.01856301 * \text{z1\_1[n]} + 0.03712602 * \text{z2\_1[n]} \\ &\quad + 0.01856301 * \text{z2\_1[n-1]} \end{aligned}$$

Second Biquad:

$$\begin{aligned} \text{z1\_2[n]} &= \text{y\_1[n]} - (-0.53945795 * \text{z1\_2[n-1]} - 0.39764934 * \text{z2\_2[n-1]}) \\ \text{z2\_2[n]} &= \text{z1\_2[n-1]}\,\text{y\_2[n]} = 1.00000000 * \text{z1\_2[n]} \end{aligned}$$

$$- 2.00000000 * z1\_2[n-1] + 1.00000000 * z2\_2[n-1]$$

Third Biquad:

$$z1\_3[n] = y\_2[n] - (0.47319594 * z1\_3[n-1] - 0.70744137 * z2\_3[n-1])$$
$$z2\_3[n] = z1\_3[n-1]\, y\_3[n] = 1.00000000 * z1\_3[n] + 2.00000000 * z1\_3[n-1] + 1.00000000 * z2\_3[n-1]$$

Fourth Biquad:

$$z1\_4[n] = y\_3[n] - (-1.00211112 * z1\_4[n-1] - 0.74520226 * z2\_4[n-1])$$
$$z2\_4[n] = z1\_4[n-1]\, y\_4[n] = 1.00000000 * z1\_4[n] - 2.00000000 * z1\_4[n-1] + 1.00000000 * z2\_4[n-1]$$

where:

- $z1_i[n]$ and $z2_i[n]$ are the state variables for the i-th biquad at time n[11]
- $x[n]$ is the input at time n[11]
- $y_i[n]$ is the output of the i-th biquad at time n[11].

This set of equations describes the mathematical expression for the given Band-Pass Butterworth IIR digital filter implemented in the provided code. Each biquad contributes to the overall filter response, and the output of one biquad serves as the input to the next biquad in the chain [28, 29, 32–34].

5 Testing and Results

5.1 Simulation of BPF in MATLAB

Here we have simulated the Band-Pass Butterworth IIR digital filter in MATLAB. First we have shown the unfiltered EMG signal acquired from the hand which consists of different frequencies (45, 60, 90, 100 and 160 Hz) and noise (Fig. 15).

And here we have shown the EMG signal after passing through the Band-Pass Butterworth IIR digital filter (Fig. 16).

And this is the frequency response of the Band-Pass Butterworth IIR digital filter with −3db point which has a sampling rate: 500.0 Hz, [4] cutoff frequency: (74.5, 149.5]) Hz [28, 29, 33, 34] (Fig. 17).

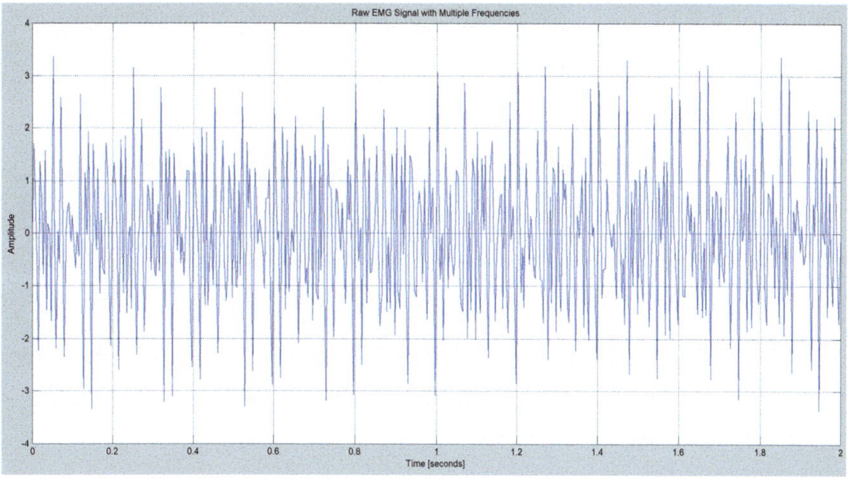

Fig. 15 Raw EMG signal with multiple frequencies of 45, 60, 90, 100 and 160 Hz [28, 29, 33, 34]

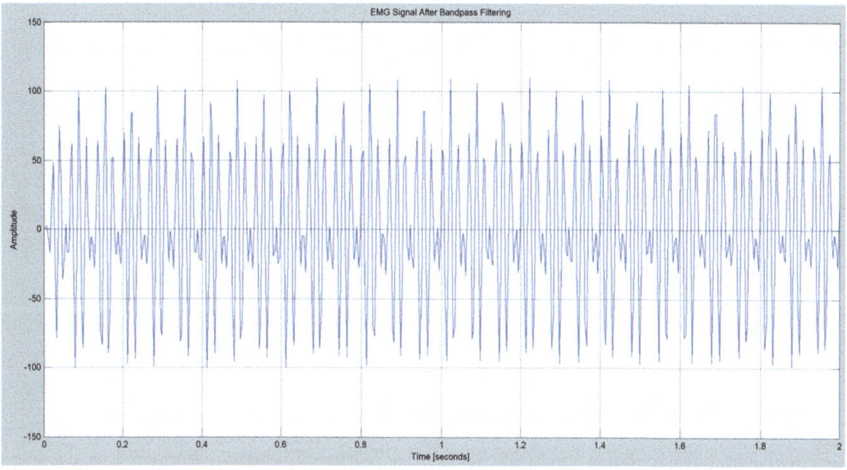

Fig. 16 Filtered EMG signal after band pass filter consisting of frequencies 90 and 100 Hz [28, 29, 33, 34]

5.2 EMG Signal, Detected Envelope

See Fig. 18.

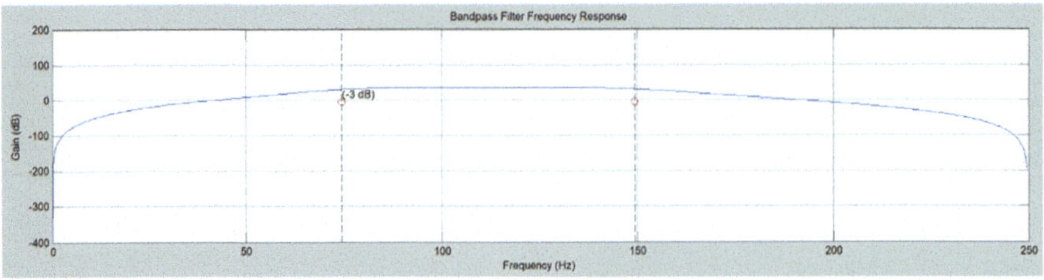

Fig. 17 Frequency response of the band-pass butterworth IIR digital filter with −3db point [28, 29, 33–35]

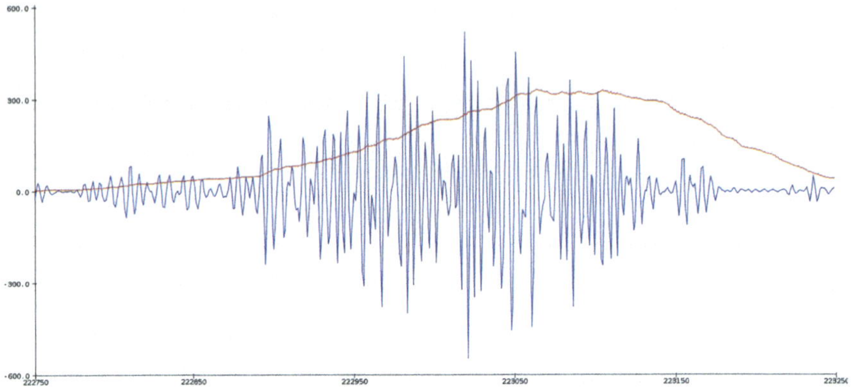

Fig. 18 EMG signal after filter with the detected envelope [2]

5.3 Testing of the Prosthetic Hand

We tested the prosthetic hand with 6 people who have healthy hands and we got these accurate results. We tested 3 things here first how accurately it can grab an object then how accurately it can close and open the hand.

Here are some images that show how the prosthetic hand looks when at rest and flexed (Figs. 19 and 20).

To test the accuracy of grabbing any object we have used a round object and a cylindrical object as shown in Figs. 21 and 22.

Now here are the results that we got after testing it on 6 people with healthy hands (Figs. 23 and 24).

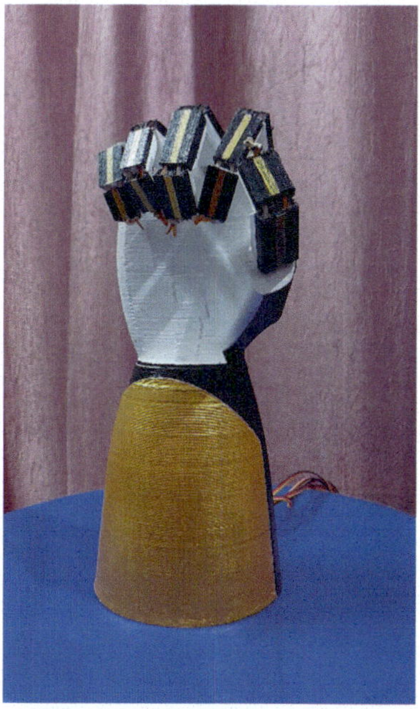

Fig. 19 Hand at flexed (close) position

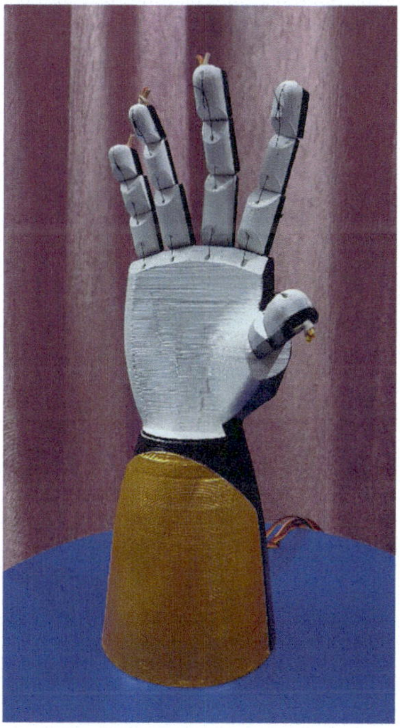

Fig. 20 Hand at rest (open) position

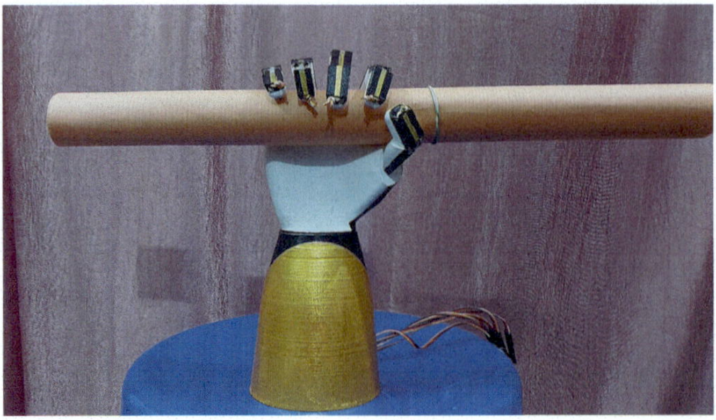

Fig. 21 Grabbing one cylindrical object

Fig. 22 Grabbing one round object

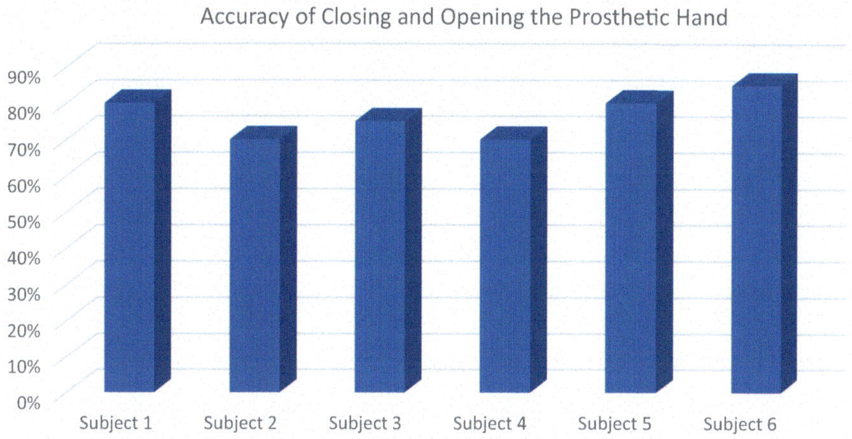

Fig. 23 Accuracy of closing and opening the prosthetic hand

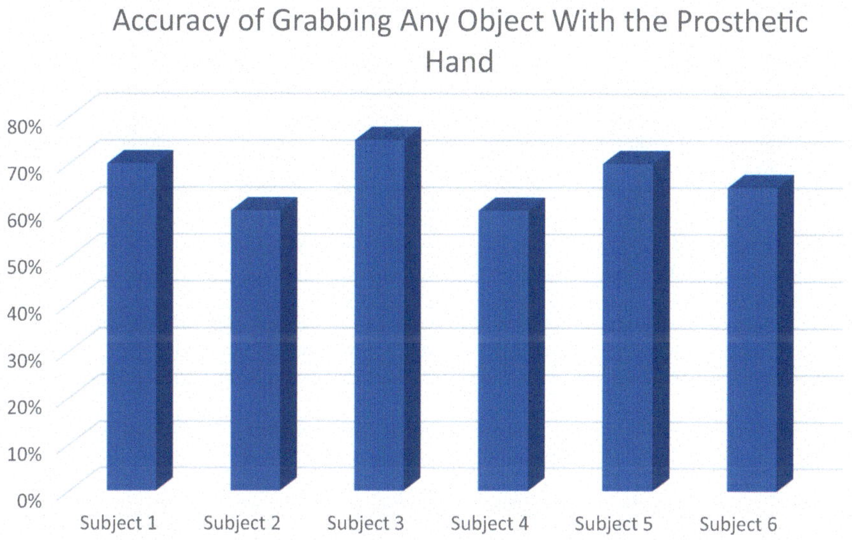

Fig. 24 Accuracy of grabbing any object with the prosthetic hand

6 Conclusion

In conclusion, the research presents a momentous advancement in prosthetic technology, particularly in the development of a bionic arm controlled by myoelectric signals. By prioritizing affordability and usability, the team aims to provide a significant solution for individuals with upper limb disabilities, especially in third-world countries where traditional prosthetics are often financially inaccessible for majority.

The unifying of noise filtering techniques, such as the Butterworth digital filter, enhances the reliability and effectiveness of the prosthetic device vastly, ensuring precise signal capture and minimizing external interference. Furthermore, the addition of a manually operated feature for holding a pen offers users greater confidence in writing. Overall, this research holds promise for improving the quality of life for amputees by offering an inexpensive and functional prosthetic solution.

References

1. Chu J-U, Moon I, Lee Y-J et al (2007) A supervised feature-projection-based real-time EMG pattern recognition for multifunction myoelectric hand control. IEEE/ASME Trans Mechatron 12:282–290. https://doi.org/10.1109/tmech.2007.897262
2. Garg S, Rathi K, Bansal AK (2023) Development of 3D printed bionic arm controlled by EMG sensor. SKIT Res J 13:13–18. https://doi.org/10.47904/ijskit.13.1.2023.13-18
3. A report on amputees in India: O&P virtual library. https://www.oandplibrary.org/op/1986_01_016.asp. Accessed 13 May 2024
4. Desk GW, Ani, Desk GV et al (2023) Home. In: Greater Kashmir. https://www.greaterkashmir.com/. Accessed 13 May 2024
5. Journal of Cardiovascular Disease Research. https://jcdronline.org/. Accessed 13 May 2024
6. Surya S, Ramamoorthy S. EMG sensor based control strategy for hand function movements using 3D prosthetic hand
7. Articles. In: IoT Design Pro—IoT Projects, Articles and News. https://iotdesignpro.com/articles. Accessed 8 May 2024
8. Upsidedownlabs Upsidedownlabs/muscle-bioamp-candy: a candy size electromyography (EMG) sensor for precise muscle bio-potential signals recording. In: GitHub. https://github.com/upsidedownlabs/Muscle-BioAmp-Candy. Accessed 5 May 2024
9. Muscle Bioamp Band (EMG band). In: Upside down labs store. https://store.upsidedownlabs.tech/product/muscle-bioamp-band. Accessed 13 May 2024
10. In: Arduino nano—soldered (without cable)—QuartzComponents. https://quartzcomponents.com/products/arduino-nano-v3-0-soldered-without-cable. Accessed 5 May 2024
11. Electronics Components Online in India. In: QuartzComponents. https://quartzcomponents.com/. Accessed 13 May 2024
12. Guest Tips to get start upload ideas and publishing resources—Pdf free download. In: docobook.com. https://docobook.com/. Accessed 13 May 2024
13. In: SG90 servo motor datasheet—components101. https://components101.com/sites/default/files/component_datasheet/SG90%20Servo%20Motor%20Datasheet.pdf. Accessed 5 May 2024
14. The ECU. In: The ScholarShip at ECU. https://thescholarship.ecu.edu/. Accessed 13 May 2024
15. Servo Motor-Understanding the basics of servo motor working. In: Servo motor basics, working principle and theory. https://circuitdigest.com/article/servo-motor-working-and-basics. Accessed 13 May 2024
16. Reconnect marvel spiderman 10000mah powerbank, Li-Po Battery, 1 USBB output port, Micro USB & Type C dual input, BIS approved, mobile accessories-DPB102 SM Online at best prices in India—jiomart. In: JioMart. https://www.jiomart.com/p/electronics/reconnect-marvel-spiderman-10000mah-powerbank-li-po-battery-1-usbb-output-port-micro-usb-type-c-dual-input-bis-approved-mobile-accessories-dpb102-sm/590041746. Accessed 5 May 2024
17. Gao G, Xu F, Xu J, Liu Z (2022) Study of material color influences on mechanical characteristics of fused deposition modeling parts. Materials 15:7039. https://doi.org/10.3390/ma15197039
18. In: National Center for Biotechnology Information. https://www.ncbi.nlm.nih.gov/. Accessed 13 May 2024

19. Hsueh M-H, Lai C-J, Wang S-H et al (2021) Effect of printing parameters on the thermal and mechanical properties of 3D-printed PLA and PETG, using fused deposition modeling. Polymers 13:1758. https://doi.org/10.3390/polym13111758
20. MDPI journal list. In: MDPI. https://www.mdpi.com/about/journals. Accessed 8 May 2024
21. Sakib N, Islam MK (2019) Design and implementation of an EMG controlled 3D printed prosthetic arm. In: 2019 IEEE international conference on biomedical engineering, computer and information technology for health (BECITHCON). https://doi.org/10.1109/becithcon48839.2019.9063190.
22. Sakib N, Islam MK (2019) Design and implementation of an EMG controlled 3D printed prosthetic arm. In: 2019 IEEE international conference on biomedical engineering, computer and information technology for health (BECITHCON). https://doi.org/10.1109/becithcon48839.2019.9063190
23. Akinci M, Ay S, Erçetin Ö (2004) Surgical treatment of macrodactyly in older children and adults. J Hand Surg 29:1010–1019. https://doi.org/10.1016/j.jhsa.2004.07.004
24. Hindawi Home. In: Hindawi. https://www.hindawi.com/. Accessed 13 May 2024
25. Chu JU, Jung DH, Lee YJ (2008) Design and control of a multifunction myoelectric hand with new adaptive grasping and self-locking mechanisms. In: 2008 IEEE international conference on robotics and automation. https://doi.org/10.1109/robot.2008.4543294
26. Enjoy free comfortable tools to publish, exchange, and share any kind of documents online! In: DocPlayer. https://docplayer.net/. Accessed 8 May 2024
27. Oaktrust Home. In: OAKTrust Home. https://oaktrust.library.tamu.edu/. Accessed 13 May 2024
28. Scipy.signal.butter#. In: scipy.signal.butter—SciPy v1.13.0 Manual. video https://docs.scipy.org/doc/scipy/reference/generated/scipy.signal.butter.html. Accessed 5 May 2024
29. Iyer SV (2021) Digital filter design using python for power engineering applications: an open source guide. Springer, S.l.
30. Brings together new 2024 papers. In: ebin.pub. https://ebin.pub/. Accessed 13 May 2024
31. In: Ulster University. https://pure.ulster.ac.uk/. Accessed 13 May 2024
32. Li Q, Jiang M, Luo B et al (2014) Design of fast second-order IIR low-pass filter based on parameter optimization. Appl Mech Mater 644–650:4382–4386. https://doi.org/10.4028/www.scientific.net/amm.644-650.4382
33. Verma NK, Ghosh AK (2019) Computational intelligence: theories, applications and future directions—volume I, ICCI-2017. Springer Singapore, Singapore
34. (2024) Digital Filter. In: Wikipedia. https://en.wikipedia.org/wiki/Digital_filter#Direct_form_II. Accessed 13 May 2024
35. Thyagarajan KS (2018) Digital filter structures. In: Introduction to digital signal processing using matlab with application to digital communications, pp 313–383. https://doi.org/10.1007/978-3-319-76029-2_8

Chapter 11
NDN: A Paradigm Shift for Scalable and Secure Data Delivery

Md. Ariful Islam, Sharmin Sultana Mim, and Partha Mandal

Abstract There is an urgent need for creative solutions to tackle the increasing problems of scalability, security, and effective content delivery caused by the tremendous strain that the fast expansion of internet connectivity has put on current infrastructure. A game-changing technology that could alter the face of online communication is named Data Networking (NDN). Unlike the traditional method, which is based only on the actual locations of data, NDN represents a paradigm change by placing an emphasis on content rather than location. An essential component of NDN's design is this change; the system is based on "data names," which enable users to directly request particular data regardless of its location. Essentially, NDN empowers content-focused routing protocols and efficient caching methods by introducing a new way of thinking about data retrieval and routing. Network function virtualization (NDN) greatly improves scalability and general network performance by placing content priority over location, therefore reducing network congestion. Additionally, NDN incorporates built-in techniques for automatic authentication of requests and responses, further integrating security measures into its design. By eliminating the need for an IP-based paradigm, this inherent security architecture allays long-standing fears about data leaks and cyberattacks while also ensuring increased privacy. NDN's architecture is highly adaptable, which allows it to handle upcoming technologies like the Internet of Things (IoT) and edge computing. The adaptable architecture of NDN makes it a vital component of the constantly changing digital world, as it effortlessly handles dynamic network conditions. Ongoing research efforts are devoted to creating and perfecting performance evaluation tools that are specific to NDN's design, with the goal of further improving its efficiency and dependability. These endeavors are centered on making NDN as effective as possible and preparing it to be easily integrated into real-world applications in different sectors. The shift to NDN isn't without its problems, but there are continuing research efforts that, like the one described here, provide light on how NDN might mould an internet ecosystem that is more safe, user-centric, and resilient. As NDN develops further, its potential to propel future improvements in digital communication and networking

Md. A. Islam (✉) · S. S. Mim · P. Mandal
Department of Electrical and Electronic Engineering, Faridpur Engineering College, Faridpur, Bangladesh
e-mail: arifular85@gmail.com

A. Acharyya et al. (eds.), *Real-World Applications and Implementations of IoT*, Studies in Smart Technologies, https://doi.org/10.1007/978-981-97-8627-5_11

highlights its importance as a game-changing change in the internet's architecture. In the end, NDN opens the door to a completely revamped internet experience, marked by increased accessibility, security, and efficiency for users all around the globe.

Keywords NDN · Named data · Forwarding information base · Interest packets · Pending interest table · Data packets · Content store · Signature based data integrity · Name-base access control

1 Introduction

Once a novelty, the internet has since evolved into the sustenance of the digital age. An abundance of data is being generated by the expanding number of connected devices, including smartphones and smart dwellings. The proliferation of data places significant demands on conventional networking techniques, resulting in lethargic performance, susceptibilities to security breaches, and challenges in obtaining necessary information.

Named Data Networking (NDN) presents itself as an innovative resolution at this juncture. NDN disregards cumbersome location-based systems in favor of data prioritization. This paradigmatic change fundamentally alters the course of our online information interactions.

Consider a library in which navigating through interminable aisles is unnecessary. Conversely, by merely specifying the book title (referred to as the data name), the library expeditiously provides the requested item. That precisely describes NDN. Data is requested by users based on its distinct name rather than its geographical coordinates. This facilitates enhanced caching, routing, and delivery capabilities, which notably mitigate network congestion and streamline the process of retrieving information [1].

NDN communication is predicated on a "conversation" that takes place between data providers and consumers. Users transmit "Interest packets" that include the names of the desired data. In their traversal of the network, these packets utilize routing information and content identifiers to determine the optimal path. After receiving an Interest packet that corresponds to its content, a data source (producer) returns a "Data packet" that includes the requested information. By performing this graceful dance, network congestion is reduced and data delivery is optimized.

Security is an additional domain in which NDN excels. NDN, in contrast to conventional networks, incorporates robust security mechanisms into its very fabric. The process of automatically authenticating requests and responses serves to protect data against unauthorized access or alteration. This security feature provides an additional level of protection in our ever more interconnected society [2].

NDN is adaptable to the most recent developments. The adaptable design of the system effortlessly accommodates emergent technologies such as edge computing and the Internet of Things (IoT) [3]. As a result, NDN is a significant participant in the continuously changing digital environment, enabling streamlined data transfer across a wide range of devices and platforms.

While transitioning to NDN can be advantageous and difficult, it also presents obstacles. Although substantial financial investments may be required, the potential benefits are immense. This investigation examines the fundamental tenets, structure, and implementations of NDN, demonstrating its capacity for profound change. We can lay the foundation for a more robust, secure, and user-centric internet experience by adopting NDN. This transition holds the potential to unveil fresh avenues for innovation and redefine the way in which we obtain information on a global scale.

2 NDN Architecture

NDN identifies data using identifiers such as file paths rather than addresses. This enables you to request the desired information (data) rather than specifying the location of the information (a device). Similar to intelligent mailboxes, NDN routers can store and distribute frequently accessed information to increase performance. There are two primary participants: publishers, who upload data bearing their label to the network; and requesters, who employ special communications to request data by its label. Routers either forward the message in the correct direction or deliver the data from their cache as these communications traverse the network [4]. NDN, in summary, places emphasis on the content itself, wherein caching nodes collaborate to expedite the delivery of the required resources (Fig. 1).

The Named Data Networking (NDN) architecture signifies a paradigmatic transition in the configuration of internet communication, substituting a content-centric approach for the conventional host-centric model [5]. Named Data Networking

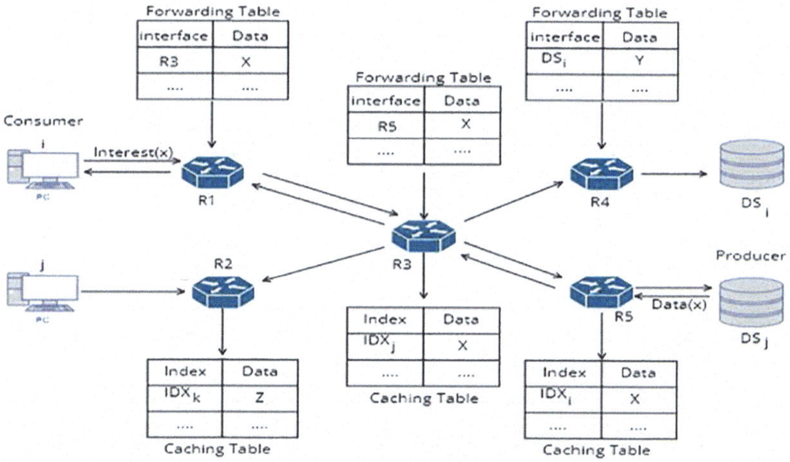

Fig. 1 Architecture of NDN

(NDN) substitutes named data for IP addresses as the primary means of communication. A conceptual overview of the NDN architecture is as follows:

a. **Named Data**: Data is the foundation of communication in NDN. Instead of employing IP addresses to identify devices (hosts), NDN prioritizes the distinctive designation of data objects [6]. A hierarchical and human-readable name is allocated to each data point, enabling users and applications to explicitly indicate their preference for particular content.
b. **Packet Structure and Data Naming**: NDN implements a hierarchical naming scheme to designate data objects. Similar to URLs, names are organized with labels being separated by slashes (/). As an illustration, "ndn://example.com/documents/research-paper" could be a name [7]. Data is identified and retrieved from the network using names.
c. **Communication Centric on Data**: In Network-Defined Networking (NDN), communication is governed by Named Data rather than host-to-host interactions. Data packets, alternatively referred to as Interest and Data packets, are utilized to transmit the respective data objects and indicate interest in specific identified entities. Interest packets, which comprise the requested data, are transmitted by producers in response to Interest packets, which are generated by consumers expressing interest in a specific piece of data.
d. **Content forwarding and routing**: The utilization of content forwarding and routing in Network-Defined Networking (NDN) is an innovative methodology based on named data. In an NDN network, routers keep Forwarding Information Base (FIB) tables that associate name prefixes with next-hop interfaces [8]. Upon receiving an Interest packet, a router consults the FIB in order to ascertain which outgoing interfaces are suitable for forwarding the Interest. Pending Interest Tables (PITs) are a type of stateful forwarding utilized by routers to monitor pending interests and match incoming data packets to satiate those interests [9].
e. **Data Caching and Retrieval**: One inherent characteristic of Network-Defined Networking (NDN) is its capacity to store and retrieve data. Consumer devices and Network Device Network (NDN) routers possess the ability to cache data objects they receive, allowing them to locally meet later queries for the same data. The implementation of this caching strategy leads to an overall enhancement in network performance, a reduction in bandwidth usage, and an improvement in efficiency.
f. **Ensuring Security and Trustworthiness**: NDN utilizes strong security techniques to protect the genuineness and consistency of data. The implementation of producer signatures on data items within Network Device Networks (NDN) enables consumers to verify the origin and reliability of the data they import. Implementing security-by-default methods enhances data reliability and mitigates potential security vulnerabilities in IP-based networks.
g. **Access Control by Name**: NDN offers precise access control based on data names. Access restrictions can be enforced by producers through the specification of access policies that are linked to data names. In order to retrieve data objects,

consumers must either possess the requisite credentials or satisfy the access requirements specified by the data producers.

The NDN architecture presents several benefits by embracing a content-centric approach: streamlined content distribution, built-in data caching, bolstered security measures, and compatibility with dynamic environments and emerging technologies. A comprehensive comprehension of the NDN architecture's conceptual overview serves as a fundamental basis for subsequent investigation and evaluation of its advantages and difficulties.

3 Essential Components and Fundamental Principles

a. **Data Names**: The data names in NDN are assigned as hierarchical and easily understandable identifiers for data objects. Names have a structure that resembles URLs, with multiple labels separated by slashes (/) [6]. Data names play a crucial role in the network by acting as distinct identifiers for content. They enable consumers to express their preferences for particular data objects.
b. **Interest Packets**: Consumers in NDN can send interest packets to indicate their desire to retrieve specific data. A customary interest packet comprises the designation of the intended data entity. Upon receipt of an Interest packet, a router proceeds to examine its Forwarding Information Base (FIB) in order to determine the appropriate outgoing interface (s) for the purpose of forwarding the Interest toward the data producer.
c. **Forwarding Information Base (FIB)**: NDN routers maintain the Forwarding Information Base (FIB), which is a routing database. Associating name prefixes with next-hop interfaces enables routers to determine the correct outgoing interface (s) for forwarding Interests toward the data producers, similar to how an IT project manager coordinates tasks and resources. The FIB can be populated either through routing protocols or by manual configuration.
d. **Pending Interest Table (PIT)**: The Pending Interest Table (PIT) is a data structure that is actively managed by NDN routers. The system keeps track of pending requests that have been forwarded but have not yet been fulfilled by receiving the necessary data packets. Upon the arrival of a data packet, the router employs the Packet Information Tree (PIT) to determine the interfaces that will be used to forward the packet. The process involves aligning the package with the pertinent pending interests.
e. **Content Store (CS)**: The Content Store (CS) serves as a cache for storing recently received Data packets in NDN routers and consumer devices. Efficient data retrieval is made possible by satisfying subsequent Interest packets locally, eliminating the need to fetch data from remote sources. The CS is managed using a content-based eviction policy, where less frequently accessed data is replaced by newer data when the cache reaches its capacity limit.

In NDN, data objects are signed by producers using cryptographic signatures to ensure data integrity and authentication. This approach guarantees the integrity and authenticity of the data. Consumers have the ability to ensure the trustworthiness of received data packets by validating the signatures using the public key of the corresponding producer.

NDN offers access control mechanisms that are based on data names, allowing for precise control over access. Access policies can be enforced by specifying rules associated with data names. Users are only able to access data objects if they have the required credentials or meet the access requirements set by the data producers.

The NDN architecture is built upon key components and principles such as data names, Interest and Data packets, FIB, PIT, CS. These elements serve as the foundation for NDN. Having a clear understanding of these components is essential in order to fully grasp how NDN works in content-centric communication and data retrieval, and to appreciate its advantages.

4 Data Naming and Packet Structure

Named Data Networking (NDN) utilizes a distinct naming scheme and packet structure that sets it apart from the traditional IP-based network architecture. Allow me to provide you with a comprehensive overview of data naming and packet structure in NDN:

a. **Data Naming**: NDN utilizes a hierarchical and user-friendly naming system to identify data objects. These names play a crucial role in identifying content in the network and making it easier to find and share data. NDN names have a structure that resembles URLs (Uniform Resource Locators) and are made up of multiple labels separated by slashes (/). As an IT project manager, you might consider using a data name like "ndn://example.com/documents/research-paper" [10].
b. **Packet Structure**: Names in the NDN system can have different levels of granularity. They have the ability to represent various types of data, including individual objects, groups of objects, and larger categories. The hierarchical structure of the names allows for effective routing and organization of data within the network. There are two main types of packets involved in NDN communication: Interest and Data packets. These packets are crucial for the request-response mechanism in NDN.
   In NDN, data packets are signed by the data producers to guarantee their integrity and authenticity. The digital signature allows users to easily confirm the authenticity and source of the data they receive.
c. **Data Packets**: Data packets are transmitted by data producers in response to received Interest packets. Just as an IT project manager would do, a data packet efficiently carries the requested data object, along with its associated name and other important metadata. Producers sign data packets to guarantee the integrity and authenticity of the data. When a router receives a Data packet, it checks

for any pending Interests in its Pending Interest Table (PIT) and sends the Data packet back along the reverse path to fulfill those pending Interests.
Data packets can contain additional metadata, such as version numbers, freshness information, and caching directives, alongside the name, payload, and signature. Efficient data caching and retrieval within the network is facilitated by this metadata.

d. **Packet Transmission**: NDN utilizes the hop-by-hop communication model to transmit Interests and Data packets. Network routers have a vital function in directing Interests to data producers and sending data packets to consumers.

When a router receives an Interest packet, it consults its Forwarding Information Base (FIB) to find the correct outgoing interface (s) for forwarding the Interest. As an IT project manager, it's important to ensure that the router maintains a Pending Interest Table (PIT) to effectively track pending Interests and match incoming Data packets to satisfy those Interests.

Upon receiving a data packet, the router employs the Packet Information Tree (PIT) to determine the interfaces that will be used for forwarding the message and addressing any outstanding interests. This particular forwarding technique facilitates the redirection of data packets via the opposite route, ultimately reaching the end users.

5 Routing and Forwarding Mechanism

Named Data Networking (NDN) presents an alternative paradigm for routing and forwarding mechanisms, distinct from the conventional IP-based routing. The routing and forwarding mechanisms of NDN are specifically engineered to convey data in an efficient manner, utilizing named content as opposed to host-centric addresses (Fig. 2).

The following is a synopsis of the forwarding and routing mechanisms utilized by NDN:

a. **FIB**: The Forwarding Information Base (FIB) is a routing table that is managed by Network Device Network (NDN) routers. It serves the purpose of linking name prefixes with next-hop interfaces. The FIB consists of entries that link specific name prefixes to the outward interfaces for interest packet forwarding. Routers employ the Forwarding Interest Block (FIB), which is populated through either human configuration or routing protocols, in order to determine the destinations for forwarding Interests.
b. **Interest Forwarding**: When a router receives an Interest packet, it examines the name in the packet and then performs a longest-prefix match query in the FIB. The router ensures that the Interest is routed through the most accurate matching prefix in the FIB by using the longest-prefix match. The selection of outgoing interfaces for forwarding the Interest packet is determined by the router, as stated by the FIB entry.

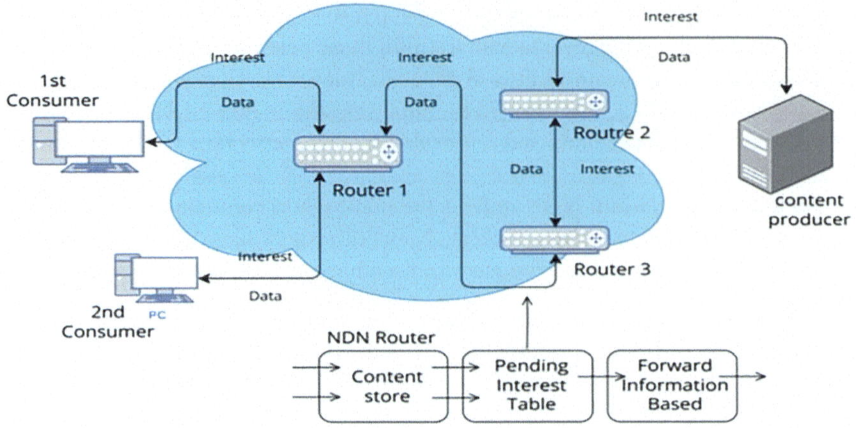

Fig. 2 Routing and forwarding mechanism of NDN

c. **Pending Interest Table (PIT)**: In NDN, routers maintain a Pending Interest Table (PIT) to track the interests they have transmitted but have not yet received the matching data packets. The PIT maintains a log of pending Interests and the incoming interfaces from which they originated. Hence, the router possesses the ability to establish a correlation between incoming data packets and the matching pertinent interests that are now present in the PIT.
d. **Data forwarding**: During the data forwarding process, a router consults its Packet Information Tree (PIT) to determine the appropriate interfaces for forwarding the data packet upon its arrival at the router. The router analyzes the Packet Information Tree (PIT) to identify entries that match the name specified in the data packet. The router initiates the transmission of the Data packet on the incoming interfaces that are linked to the appropriate pending Interests. The process known as reverse path forwarding involves redirecting the data payload via the opposite path of the interest in question.
e. **Interest Retransmission**: If an Interest packet in the PIT times out without getting a corresponding Data packet, routers have the capability to perform interest retransmissions. The routers retransmit the Interests on the outgoing interfaces that are connected to the pending Interests in the PIT. Through the utilization of this method, stakeholders are provided with several opportunities to establish communication with the data producer and acquire the relevant data payload.
f. **Adaptive Forwarding**: NDN routers can employ adaptive forwarding techniques based on local caching and content prominence. Based on prior observations, routers have the ability to prioritize the forwarding of Interests toward content that is more likely to be available in the local Content Store (CS) or has experienced higher levels of popularity. Adaptive forwarding algorithms not only enhance network efficiency but also reduce superfluous traffic.

g. **Multicast Support**: NDN inherently incorporates multicast communication. NDN facilitates the consolidation of multiple data sources or producers into a single Interest bundle. Interests can be duplicated by routers and forwarded to multiple sources. Upon arrival, data packets are transmitted to all interfaces that are linked to the corresponding pending interests in the PIT. This process facilitates the efficient delivery of multicast data.

Through the utilization of adaptive forwarding, FIB, PIT, interest forwarding, data forwarding, interest retransmission, and multicast support, the routing and forwarding mechanisms of NDN facilitate content-centric communication in an efficient and scalable manner. These mechanisms facilitate the provision of designated data objects to consumers in accordance with their expressed preferences and the network's availability of corresponding data.

6 Advantages of NDN

a. **Content-Centric Approach**: NDN turns communication from being host-centric to being content-centric. Specifically, NDN emphasizes the naming and retrieval of data objects directly, rather than relying on host locations and IP addresses. Enhanced data caching, better scalability, and effective content distribution are just a few benefits of this content-centric strategy.
b. **Efficient Content Retrieval**: Name-based queries rather than location-based queries are used to get data in NDN. Customers will no longer need to be aware of the precise addresses of content providers thanks to this. Customers can access material from any source—be it the original creator or a network-cached copy by expressing their interests in named data items. Efficient content retrieval is made possible by this flexibility and the separation of material from particular hosts.
c. **Data Caching and Localized Content Delivery**: Consumer devices and NDN routers have the ability to store received data items in caches. The caching method minimizes the need for repetitive data transfer over the network by enabling local fulfillment of subsequent requests for the same data. Caching makes it easier to serve localized content while using less bandwidth and speeding up response times, especially for popular content.
d. **Enhanced Scalability**: The content-centric design of NDN provides benefits for scalability from the outset. NDN makes it possible for material to be distributed around the network more effectively by doing away with the need for host addresses. The network may grow to support a huge number of data objects and users thanks to the hierarchical naming strategy, which makes routing and forwarding based on name prefixes efficient and less difficult.
e. **Data Integrity and Security**: NDN has built-in security features. The producers of the data objects in NDN sign them, guaranteeing the authenticity and integrity of the data. By authenticating the digital signatures, users can confirm the

authenticity and integrity of the material they have received. This security-by-default architecture improves data trustworthiness while reducing security flaws in IP-based networks.

f. **Name-based Access Control**: NDN provides data names as the basis for fine-grained access control. By defining access policies linked to certain data names, producers can apply restrictions on access. This makes it possible to regulate access to particular data objects more precisely. Name-based access control improves the security and privacy of data.
g. **Support for Dynamic Environments**: Network topology and content are subject to frequent changes in dynamic environments, which is where NDN excels. Content objects can change in position or presence after they are identified, but this won't impact their retrieval. Because of its increased resilience to mobility, dynamic content availability, and network disturbances, NDN is well-suited for new applications and mobile settings.
h. **Integrated Support for Multicast**: Multicast communication is supported by NDN by default. Several data sources or producers can satisfy a single Interest packet, enabling effective multicast data distribution. This is especially helpful in situations where data needs to be distributed to several receivers at once.

7 Challenges and Limitations

a. **Transition from IP-based Networks**: One of the main obstacles is making the switch from the current IP-based network infrastructure to NDN. There are two limitations and challenges associated with this. NDN is a paradigm shift in networking that calls for significant adjustments to applications, protocols, and network equipment. During the transition period, NDN's coexistence and interoperability with legacy systems present issues.
b. **Naming and Scalability**: Despite the flexibility and effective routing offered by the hierarchical naming strategy in NDN, maintaining a scalable and globally unique naming system can be difficult. Ensuring the uniqueness of names, upholding naming norms, and averting naming disputes become challenging jobs as the quantity of data items and users increases.
c. **Routing Efficiency**: For the best content retrieval in NDN, routing must be done efficiently. It can be difficult to create scalable and effective routing protocols that can manage a lot of Interests and Data packets in dynamic network environments. As a network grows in size, maintaining routing tables and making forwarding decisions becomes more complex, necessitating careful design and optimization.
d. **Data Freshness and Staleness**: Caching is essential to the delivery of material across NDNs. It can be difficult to guarantee data freshness and avoid data staleness in caches, nevertheless. Data items might change over time, thus it's critical to keep caches updated to prevent giving users stale or obsolete information.

e. **Overhead and Resource Utilization**: Compared to IP-based networks, NDN adds more overhead and packet processing. Computational and storage needs rise with the requirement for signature verification, Interest/Data packet matching, and content-based routing decisions. For wider adoption, there is a challenge of maximizing device and network resources with the least amount of overhead.
f. **Adoption and Deployment**: Industry backing, community agreement, and standardization initiatives are necessary for the effective implementation and broad acceptance of NDN. To fully utilize NDN, it will be necessary to overcome the inertia of current network infrastructure, implement NDN-compatible hardware and software, and create an ecosystem of services and applications.

The wider acceptance and use of Named Data Networking depends on addressing these obstacles and restrictions through research, development, and cooperation among academic institutions, business, and standards organizations.

8 Comparison with Traditional IP-Based Internet Architecture

NDN has benefits including faster content retrieval, data caching, and increased scalability, but it also has drawbacks with regard to naming, routing, transitioning, and interoperability with current systems. Although traditional IP-based networks have a robust ecosystem of applications and protocols and a well-established infrastructure, they may have difficulties with content-centric communication, effective caching, and multicast support. The selection between Network-Defined Networking (NDN) and traditional IP-based networks is contingent upon the specific requirements, attributes, and qualities of the network infrastructure and applications (Table 1).

9 Exploring the Practical Applications of NDN

The content-centric approach of Named Data Networking (NDN) brings about a significant transformation in traditional data communication. NDN prioritizes the transmission of content over the physical location of data, in contrast to typical network topologies that primarily concentrate on data placement. The utilization of caching methods enables the availability of a wide range of practical applications. Enabling edge computing, NDN enhances localized data processing, optimizes sensor data management for smart cities, and improves data retrieval and security in the Internet of Things (IoT). Additionally, Network-Defined Networking (NDN) improves the distribution and caching of content, hence enhancing the speed of content delivery. Furthermore, NDN provides support for secure and context-aware services. All of these aspects play a crucial role in providing a robust foundation for data-driven applications (Fig. 3).

Table 1 NDN versus IP based internet architecture

SI no	Area	NDN	IP based network
1.	Focus	Communication that focuses on the content. NDN prioritizes the direct naming and retrieval of data objects, irrespective of their location or host.	Communication focused on the host. IP networks are dependent on the utilization of host IP addresses for addressing and routing purposes.
2	Naming and addresses	Content is identified using hierarchical and easily understandable data names. Names are not influenced by geographical location and have the ability to convey meaning and adapt	IP addresses serve the purpose of verifying the identity of hosts and network interfaces. IP addresses are contingent on geography and have limited expressiveness
3	Data retrieval	The expression of consumer interests in named data objects is facilitated by routers, which then proceed to locate and retrieve the associated data objects	Communication is initiated by consumers through the transmission of packets to designated destination IP addresses. Routing tables are utilized by routers to transmit packets according to IP addresses
4	Routing	Routers decide the next-hop interfaces for Interest packet forwarding based on names and longest-prefix matching. The routing of data packets is determined by the Pending Interest Table (PIT), which contains information about pending interests	In the context of packet forwarding, routers employ IP addresses to ascertain the subsequent-hop interfaces by referencing routing tables that encompass IP address prefixes
5	Caching and data dissemination	Data objects can be cached by routers and consumer devices. Data can be obtained from several sources, such as caches, hence minimizing the necessity for redundant data transmission	The process of caching is commonly executed at the application layer. The dissemination of content occurs via direct connection between hosts
6	Security	Data integrity and authenticity are ensured through the implementation of built-in security methods, such as data object signing. Interest-based access control (IBAC) enables precise and detailed management of data access	The implementation of security measures is predominantly focused on higher layers, such as transport layer security (TLS), which is employed to ensure secure communication. IP addresses are commonly used as the basis for access control

(continued)

Table 1 (continued)

SI no	Area	NDN	IP based network
7	Multicast support	The system inherently facilitates multicast communication, allowing various data sources to satisfy a single Interest packet	The provision of multicast functionality necessitates the utilization of specialized protocols and infrastructure
8	Compatibility and transition	A shift from the current IP-based infrastructure and application environment is necessary. It is necessary to tackle the issues of coexistence and interoperability with IP-based networks	The infrastructure is extensively deployed and well-established, boasting a substantial array of pre-existing applications and protocols

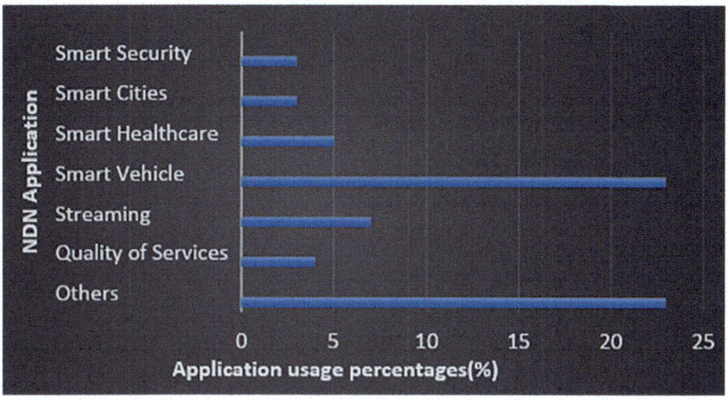

Fig. 3 Emerging patterns in the implementation of NDN

a. **Internet of Things (IoT)**: The domain of the Internet of Things (IoT) presents numerous applications and use cases where Named Data Networking (NDN) exhibits considerable potential. In order to harness the content-centric architecture of NDN, IoT deployments can benefit from efficient data retrieval, caching, and enhanced security [11]. Here are some uses of the applications of NDN in IoT:

   i. **Sensor Data Collection and Monitoring**: NDN enables efficient collection and dissemination of sensor data in IoT applications. Sensors can publish data using unique names, allowing interested parties to express interest in specific data types or sensors. Caching capabilities reduce redundant data transmission, improving availability and response times.

   ii. **Context-Aware Services**: NDN facilitates context-aware IoT services by allowing devices and applications to express Interest in specific contextual information. For example, a smart home system can express an Interest in the current room temperature, and NDN delivers the corresponding data from the nearest source, enabling dynamic and personalized services based on real-time context [12].

iii. **Secure Data Sharing and Access Control**: NDN's content-centric nature inherently supports data security and access control. Data objects can be signed by producers, ensuring integrity and authenticity. Access control policies can be enforced at the data naming level, providing fine-grained control over data access, which is crucial for secure data sharing in healthcare or industrial control systems.

iv. **IoT Data Fusion and Aggregation**: NDN facilitates data fusion and aggregation in IoT deployments with multiple data sources. Interests can be expressed for specific data types, and NDN can aggregate and deliver corresponding data from multiple sources, enabling efficient data fusion for higher-level analysis and decision-making [13].

v. **Mobility and Location-Based Services**: NDN's content-centric approach suits mobile IoT devices. Devices can maintain their unique names while moving between network locations. NDN can deliver data to mobile devices based on their names, regardless of the network attachment point, enabling seamless mobility and location-based services like asset tracking or mobile healthcare applications [14].

vi. **Energy Efficiency and Resource Management**: NDN's caching mechanisms contribute to energy efficiency in IoT deployments. Caching frequently requested data at the network edge reduces data transmission and conserves energy. NDN's naming-based routing allows devices to retrieve data from nearby sources, reducing energy consumption and network congestion.

These examples demonstrate the application of NDN in various IoT contexts. The content-centric architecture of NDN provides distinct benefits in terms of data retrieval, caching, security, and context awareness, making it a promising technology for addressing IoT application requirements.

b. **Edge Computing**: In the domain of edge computing, Named Data Networking (NDN) presents a diverse range of applications and use cases, wherein data processing and analytics take place at the network edge [15]. Edge computing installations can derive advantages such as efficient data retrieval, reduced latency, and improved scalability by leveraging NDN's content-centric design and caching methods.

I. **Distributed Data Processing**: In edge computing environments, NDN facilitates distributed data processing across edge devices and local servers. Data objects can be cached at the network edge, enabling nearby devices and applications to retrieve and process data locally. This reduces the reliance on transmitting data to centralized servers, resulting in minimized latency and reduced network congestion.

II. **Localized Analytics and Decision Making**: NDN's caching capabilities at the network edge enable localized analytics and decision-making in edge computing scenarios. Relevant data objects can be cached in proximity to specific edge devices or applications, enabling real-time processing and immediate access to required data. This is particularly valuable for time-sensitive applications like real-time monitoring, anomaly detection, or event-triggered actions.

III. **Content-Based Data Retrieval**: NDN's content-centric approach allows data retrieval based on content names rather than traditional IP addresses. In edge computing, this enables edge devices and applications to express Interest in specific data objects they require, and the NDN network can deliver the corresponding data from nearby caches or local servers [16]. This content-based data retrieval improves efficiency and reduces the burden on centralized data repositories.

IV. **Dynamic Service Provisioning**: NDN enables edge devices to express Interest in specific services they need, and the network can dynamically provision those services at the edge [17]. This facilitates on-demand service delivery, reducing reliance on centralized cloud resources. For instance, an edge device can express an interest in a particular video processing service, and the NDN network can identify and deliver the nearest available service to the requesting device.

V. **Scalable Edge Infrastructure**: NDN's caching capabilities enable efficient content distribution and data sharing across edge devices and servers. As more edge devices join the network, data objects can be cached and shared among them, reducing the necessity of transmitting data to centralized servers. This distributed caching and content sharing enhance scalability as the edge infrastructure can handle growing data volumes and demand without overwhelming the network.

VI. **Resilient and Fault-Tolerant Edge Networks**: NDN's inherent caching support enhances the resilience of edge networks in the face of failures or disruptions. If a centralized server or service becomes unavailable, edge devices can still retrieve cached data from nearby caches, ensuring uninterrupted operation. This fault-tolerant characteristic of NDN enhances the reliability and availability of edge computing systems [18].

VII. **IoT Integration at the Edge**: NDN facilitates the integration of IoT devices and data sources at the network edge. Edge devices can express Interest in specific IoT data objects, and the NDN network can deliver the corresponding data from nearby caches or IoT devices [19]. This enables localized data processing, reduces latency, and optimizes the utilization of IoT data in edge computing applications.

These use cases demonstrate how NDN can be applied in edge computing scenarios to enhance data processing, reduce latency, improve scalability, and enable localized analytics and decision-making. NDN's content-centric architecture, coupled with caching and data retrieval mechanisms, makes it a promising technology for efficient and scalable edge computing deployments.

c. **Smart Cities and Infrastructure**: One emerging technology, known as Named Data Networking (NDN), holds great potential for enhancing the efficiency and scalability of smart cities. The distinguishing characteristic of NDN, in contrast to the existing Internet architecture, is in its content-centric approach. This implies that material is designated and retrieved based on its content rather than its physical location [20]. This enables the caching of data in proximity to its required location, hence enhancing performance and minimizing latency. NDN can be used in a variety of smart city applications, including:

   I. **Sensor Data Collection and Monitoring**: NDN can be used to collect and distribute sensor data more efficiently. Sensors can publish data using unique names, and interested parties can express interest in specific data types or specific sensors. This can reduce the need for repeated data transmission and improve data availability and response times.
   II. **Real-time Traffic Management**: NDN can be used to support real-time traffic management. Traffic sensors, cameras, and other devices can generate data about traffic conditions, which can be published and cached using NDN [21]. Traffic management systems and applications can then express interest in this data, allowing them to retrieve real-time information about traffic congestion, accidents, or road conditions. This can help to improve traffic routing, congestion avoidance, and urban mobility.
   III. **Environmental Monitoring and Management**: NDN can be used to monitor and manage the environment more efficiently. Sensors that measure air quality, noise levels, temperature, and other environmental parameters can publish their data using NDN. Environmental management systems and applications can then express interest in specific environmental data, enabling real-time monitoring, analysis, and decision-making for urban sustainability and improved quality of life.
   IV. **Energy Management and Conservation**: NDN can be used to manage and conserve energy more effectively. Smart meters, energy monitoring devices, and renewable energy sources can publish energy data using NDN. This data can be cached and shared, allowing energy management systems and applications to retrieve and analyze it in real time. This can help to improve energy load balancing, demand response, and optimization of energy consumption within the city.
   V. **Public Safety and Emergency Response**: NDN can be used to enhance public safety and emergency response. Sensors, cameras, and emergency services can publish and cache data related to public safety, such as crime incidents, fire incidents, or emergency alerts. Emergency response systems and applications can then express interest in this data, enabling real-time

access to critical information for timely emergency response and improved situational awareness.

VI. **Smart Grid Management**: NDN can be used to manage the smart grid more efficiently. Energy-related data, such as electricity generation, distribution, and consumption, can be published and cached using NDN. This enables energy management systems to retrieve real-time data, perform analysis, and optimize the operation of the smart grid. NDN's caching mechanisms also facilitate localized energy data retrieval, reducing the need for backhaul communication to centralized servers.

VII. **Data Sharing and Collaboration**: NDN's content-centric approach simplifies data sharing and collaboration among different entities within a smart city. Public services, city agencies, and stakeholders can publish and cache data objects relevant to their domain, making the data easily accessible to authorized parties. This encourages data sharing, interconnectivity, and collaboration among various smart city applications and services.

These are just a few examples of how NDN can be used in smart cities. Further research, standardization, and collaboration are needed to realize the full potential of NDN in this domain.

10 Security and Privacy in NDN

a. Mechanisms for Authentication and Authorization: In the realm of networking architecture, security and privacy hold paramount importance, particularly in the context of Named Data Networking (NDN). NDN employs multiple measures to effectively mitigate these problems and guarantee the preservation of data confidentiality, integrity, and availability. The establishment of trust and management of data access in Network-Defined Networking (NDN) is heavily reliant on the implementation of authentication and authorization procedures [22]. Authentication and permission play essential roles in Network Device Networking (NDN). The key aspects of authentication and authorization in NDN are:

   i. **Data Integrity and Trust**: NDN ensures data integrity by using digital signatures that allow data consumers to verify the authenticity and integrity of data objects. This prevents unauthorized modifications and establishes trust in the data and the producer's identity.

   ii. **Certificate-Based Authentication**: NDN can utilize certificates to authenticate entities in the network. Certificates issued by trusted authorities bind public keys to entity identities, enabling data consumers to verify the signature using the producer's public key obtained from the certificate. This ensures that only trusted entities can produce or consume data within the NDN network.

iii. **Access Control**: NDN supports fine-grained access control mechanisms based on content names. Producers can define access control policies at the namespace level, determining who can access their data objects. Consumers need appropriate credentials or must satisfy specific access requirements to retrieve data, ensuring only authorized entities can access and consume data within the NDN network.

iv. **Trust Models and Trust Management**: NDN allows flexible trust models where entities establish trust relationships based on their criteria. Trust can be established through direct interactions, certificates, reputation systems, or a combination of approaches. Trust management mechanisms enable entities to evaluate and manage the trustworthiness of producers and consumers, ensuring secure and reliable data exchange.

v. **Privacy Preservation**: NDN protects privacy by decoupling data names from location and identity. Producers can assign meaningful names to data objects without revealing their physical location or identity, enhancing privacy. Encryption techniques can also be applied to data objects to ensure that only authorized consumers with decryption keys can access the data.

vi. **Secure Name Resolution**: NDN employs secure name resolution mechanisms to prevent name spoofing and ensure the integrity of routing information. Namespace Providers (NPs) are responsible for managing namespaces and resolving names to data objects. NPs can use secure protocols and authentication mechanisms to provide accurate and tamper-proof name resolution, maintaining overall network security.

These authentication and authorization mechanisms in NDN contribute to the security and privacy of the network. By ensuring data integrity, authenticating entities, implementing access control, managing trust, and preserving privacy, NDN establishes a robust security framework for data-centric networking. Ongoing research and evaluation are necessary to address emerging security challenges and evolving threat landscapes.

b. **Content Integrity and Trustworthiness**: Maintaining content integrity and trustworthiness is a critical aspect of data-centric networking, such as Named Data Networking (NDN). In NDN, it is crucial to ensure the integrity of content and establish trust in the data to uphold the reliability and security of the network [23]. The following are key considerations regarding content integrity and trustworthiness in NDN:

i. **Digital Signatures**: NDN employs digital signatures to guarantee content integrity. The signature field is present in every Data packet and serves to bind the content of the packet to its name [24]. Producers sign their data objects using private keys, creating digital signatures that consumers can verify using corresponding public keys. During data retrieval, consumers can verify the signature to ensure that the content hasn't been tampered with during transmission. Digital signatures provide strong cryptographic assurance of data integrity, ensuring the content remains intact and unaltered.

ii. **Producer–Consumer Trust**: NDN fosters trust between producers and consumers through the use of digital signatures. By verifying the digital signature, consumers can establish trust in the data's authenticity and the producer's identity. This trust relationship is crucial for consumers to rely on content from trusted sources, minimizing the risk of receiving tampered or malicious data.
iii. **Certificate-based Authentication**: In order to guarantee safe communication in NDN, it is necessary for data producers to have a minimum of one cryptographic key pair. Producers utilize their private keys for the purpose of signing data packets, while consumers employ public keys to authenticate them. Every public key is encoded in a digital certificate format known as X509, which is signed by the issuer [25]. Trusted authorities issue certificates that bind public keys to entity identities, such as producers. By validating the certificate chain and verifying the producer's digital signature using the public key from the certificate, consumers can establish trust in the producer's identity and the content's integrity. Certificate-based authentication enhances the trustworthiness of data exchanged in the NDN network.
iv. **Trust Models and Reputation Systems**: NDN supports trust models and reputation systems to assess the trustworthiness of producers and content. Trust can be established through direct interactions, certificates, reputation scores, or other criteria. In order to ascertain the authorization of a producer to generate data within a specific namespace, consumers necessitate a way to validate whether the producer's key possesses the requisite authority to sign a Data packet within said namespace [26]. These models allow consumers to evaluate the reliability, consistency, and quality of content from different producers. This empowers consumers to make informed decisions about the trustworthiness of the data they consume in the NDN network.
v. **Content Validation and Verification**: Consumers can employ content validation and verification mechanisms to ensure data trustworthiness. This can involve verifying digital signatures, comparing content with trusted sources, or applying validation rules and policies. Content validation and verification mechanisms help identify potential data tampering, corruption, or unauthorized modifications, ensuring that only trusted and reliable data is accepted and processed within the NDN network (Fig. 4).

Neural Networsk Device Networks (NDN) employ a range of techniques, including digital signatures, authentication, trust models, and content validation, in order to effectively tackle these challenges. The enhancement of security and trustworthiness of content transferred inside the NDN network necessitates ongoing study, standardization, and collaboration.

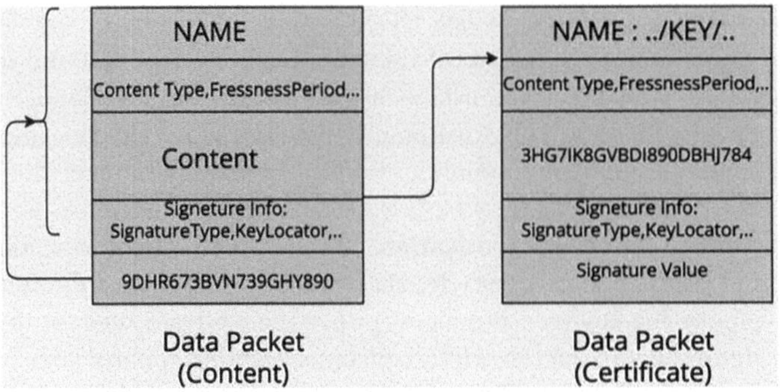

Fig. 4 Content integrity and trustworthiness of NDN

11 Implementation and Deployment of NDN

a. **Existing NDN Implementations and Frameworks**: There are several popular implementations and frameworks available for Named Data Networking (NDN), each catering to specific use cases and development requirements. Some of the notable ones include:

    I. **NDN-IoT**: NDN-IoT is a streamlined Near Field Communication (NDN) implementation specifically designed for Internet of Things (IoT) devices. The system is constructed using the Contiki operating system and provides a cost-effective and energy-efficient solution for Internet of Things (IoT) applications, with a specific emphasis on energy efficiency and sensor networks [27].

    II. **NDN-Linux**: NDN-Linux is an extensive NDN implementation specifically developed for computers based on the Linux operating system. The utilization of the ndn-cxx library facilitates the creation of a wide range of NDN applications. The NDN-Linux architecture is highly suitable for applications that necessitate high levels of power and scalability, such as content distribution networks [28].

    III. **NDN-Sim**: NDN-Sim is a specialized simulation framework designed exclusively for NDN. The platform offers a controlled environment for the purpose of testing and analyzing NDN protocols and applications. According to reference [29], NDN-Sim is a highly advantageous option for developers who aim to examine and authenticate their NDN apps.

    IV. **NDN-API**: The NDN-API is an application programming interface (API) designed specifically for the construction of Network Device Networks (NDNs). The platform provides a streamlined methodology for developing NDN applications, eliminating the need for substantial familiarity with the intricacies of the underlying protocol. The NDN-API is well-suited for developers who seek to efficiently and seamlessly construct NDN applications [30].

These implementations and frameworks offer a variety of options for developers engaged in NDN, allowing them to choose the most suitable alternative according to their own requirements. The wide range of services, including lightweight Internet of Things (IoT) devices, robust Linux systems, simulation environments, and streamlined development, effectively address different use cases and improve the efficiency and efficacy of Network Device Networking (NDN) applications.

b. **Adoption Challenges and Considerations**: Named Data Networking (NDN) is a novel networking technique designed to address the constraints of the existing Internet, including issues related to scalability, security, and privacy. However, prior to the widespread use of NDN, there are various issues that must be addressed.

   I. **Cost**: Deploying and maintaining an NDN network may entail higher costs compared to a traditional IP network.
   II. **Complexity**: The complexity of NDN surpasses that of IP, which may present challenges for certain businesses when it comes to implementing and overseeing the technology.
   III. **Interoperability**: The current lack of interoperability between NDN and the pre-existing IP network may pose challenges to its widespread adoption and seamless integration with incumbent systems.

In general, Network-Defined Networking (NDN) exhibits considerable promise as a networking architecture with the capacity to address existing difficulties within the Internet. However, it is essential to overcome the current barriers in order to achieve widespread implementation of NDN.

12 Conclusion

The concept of content-centric communication forms the foundation of Named Data Networking (NDN), a novel networking method. In Network Data Networks (NDN), data is defined and retrieved based on its name, rather than its physical location. Compared to standard IP-based networks, this offers numerous benefits such as improved content delivery, less network congestion, and heightened data security and privacy.

This study investigates the potential implications of Network-Defined Networking (NDN) on the future of networking designs. The text explores the benefits of NDN, along with the obstacles and constraints that must be overcome in order to achieve full realization of NDN. This research additionally investigates the applications and utilization scenarios of Network-Defined Networking (NDN) throughout diverse domains, encompassing the Internet of Things (IoT), edge computing, smart cities, and infrastructure.

The article places considerable emphasis on security and privacy, delving into several processes such as authentication, authorization, content integrity, trustworthiness, and privacy preservation within the context of Network Device Networks (NDN). These factors are essential in the development of robust and reliable systems that can safeguard sensitive information and uphold user confidentiality.

In summary, the study offers a thorough examination of NDN, encompassing its benefits, difficulties, and uses. This study adds to the current corpus of information on NDN and establishes a basis for future investigation and advancement in this captivating domain.

The article asserts that NDN possesses the capacity to fundamentally transform our methods of communication, information sharing, and the construction of safe and efficient networks in the forthcoming years. The authors anticipate that the publication will serve as a catalyst for additional investigation, cooperation, and advancement in the domain of Named Data Networking.

References

1. Jacobson V, Wren J, Siltanen C, Murphy A, Karthikeyan S (2011) A named data networking architecture. In: Proceedings of the 2011 ACM SIGCOMM conference on data communication, ser. SIGCOMM 11. Association for Computing Machinery, New York, NY, USA, pp 1–13
2. Taleb TA, Akkaya KN (2018) An overview of security support in named data networking. In: 2018 15th IEEE international conference on networking, sensing and control (ICNSC), pp 1–6
3. Rimac-Drobac I et al (2020) A survey on content-centric networking for edge computing. IEEE Commun Surv Tutor 22(2):1074–1101
4. Koponen T et al (2010) A practical implementation of content-centric networking. In: SIGCOMM' 10: proceedings of the 2010 ACM SIGCOMM conference on data communication, pp 183–194
5. Jacobson V et al (2009) Named data networking (NDN) project. NDN project technical report NDN-TR 0.1
6. Xylomenos G et al (2017) A survey of content-centric networking. IEEE Commun Surv Tutor 19(3):1645–1672, third quarter
7. Saxena D, Raychoudhury V (2016) Radient: scalable, memory efficient name lookup algorithm for named data networking. J Netw Comput Appl 63:1–13
8. Li F, Chen F, Wu J, Xie H (2014) Longest prefix lookup in named data networking: how fast can it be? In: Proceedings of the 2014 9th IEEE international conference on networking, architecture, and storage (NAS), Tianjin, China, pp 186–190
9. Al-qutwani M, Wang X, Yi B (2019) Name lookup in named data networking: a review. Information 10(3):85
10. Dai H, Lu J, Wang Y, Liu B (2015) BFAST: unified and scalable index for NDN forwarding architecture. In: Proceedings of the 2015 IEEE conference on computer communications (INFOCOM), San Francisco, CA, USA, pp 2290–2298
11. Zhang Z, Zhang H, Newberry E, Mastorakis S, Li Y, Afanasyev A, Zhang L (2020) Named data networking (NDN) for the internet of things (IoT). IEEE Commun Surv Tutor 22(3):1734–1757
12. Zhang Z, Zhang H, Newberry E, Mastorakis S, Li Y, Afanasyev A, Zhang L (2020) Named data networking (NDN) for context-aware IoT applications. IEEE Commun Surv Tutor 22(3):1758–1783
13. Zhang Z, Zhang H, Newberry E, Mastorakis S, Li Y, Afanasyev A, Zhang L (2020) Named data networking (NDN) for data fusion in IoT applications. IEEE Commun Surv Tutor 22(3):1784–1807

14. Zhang Y, Afanasyev A, Burke J, Zhang L (2016) A survey of mobility support in named data networking. In: Proceedings of the 2016 IEEE conference on computer communications workshops (INFOCOM WKSHPS), San Francisco, CA, USA, pp 83–88
15. Zhang Z, Zhang H, Newberry E, Mastorakis S, Li Y, Afanasyev A, Zhang L (2020) Named data networking for edge computing. IEEE Commun Surv Tutor 22(3):1808–1833
16. Huang K, Zhang R, Wang X, Zhang J (2020) Content-centric networking for edge computing: a survey. IEEE Access 8:88269–88294
17. Zhang L, Zhang H, Zhou L, Zhang Z (2021) Dynamic service provisioning in named data networking. IEEE Trans Serv Comput 14(2):255–270
18. Zhang H, Zhang L, Zhou L (2021) Fault-tolerance in named data networking. IEEE Trans Serv Comput 14(2):271–285
19. Hajibabaei MM, Habibi A, Mousavi SM (2021) NDN for IoT: a survey. IEEE Commun Surv Tutor 23(1):458–482
20. Zhang Z, Zhang H, Newberry E, Mastorakis S, Li Y, Afanasyev A, Zhang L (2020) Named data networking (NDN) for smart cities. IEEE Commun Surv Tutor 22(3):1734–1757
21. Zhang H, Zhang L, Zhou L (2020) Named data networking (NDN) for real-time traffic management. IEEE Trans Intell Transp Syst 21(6):2225–2237
22. Afanasyev A, Zhang L, Zhang H (2020) Security considerations in named data networking. IEEE Commun Surv Tutor 22(3):1758–1781
23. Zhang H, Zhang L, Zhou L (2021) Data integrity and trust in named data networking. IEEE Trans Inf Forensics Secur 16(6):1837–1852
24. Li Y, Zhang Z, Wang X, Lu E, Zhang D, Zhang L (2019) A secure sign-on protocol for smart homes over named data networking. IEEE Commun Mag 57:62–68
25. Zhang Z, Yu Y, Afanasyev A, Zhang L (2017) NDN certificate management protocol (NDNCERT). Technical report NDN-0050, NDN: Helensvale, Australia
26. Nour B, Khelifi H, Hussain R, Mastorakis S, Moungla H (2021) Access control mechanisms in named data networks: a comprehensive survey. ACM Comput Surv 54:61
27. Zhang H, Zhang L, Zhou L (2021) NDN-IoT: a lightweight NDN implementation for the internet of things. IEEE Commun Mag 59(1):122–129
28. Zhang H, Zhang L, Zhou L (2021) NDN-Linux: a comprehensive NDN implementation for Linux-based systems. IEEE Commun Mag 59(1):130–137
29. Zhang H, Zhang L, Zhou L (2021) NDN-sim: a simulation framework for named data networking. IEEE Commun Mag 59(1):138–145
30. Zhang H, Zhang L, Zhou L (2022) NDN-API: an efficient API for named data networking. ACM Trans Internet Technol 22(1):84–109

Chapter 12
Assessment of Agricultural Condition Using Composite Sensor Network for Production Improvement

Arpan Deyasi, Soumyadip Dhar, Hiranmoy Roy, and Poly Saha

Abstract A novel sensor network is developed for the detection of necessary parameters related with environmental condition which has direct influence on the growth rate of potato plant. The embedded system is efficient enough to determine the real-time condition of soil in terms of pH, water content of the soil at runtime in ppm unit, NPK (Nitrogen-Phosphorous-Potassium content), TDS (Total Dissolved Solid); and atmospheric status including temperature, humidity, and intensity of Sun. Periodical growth monitoring is performed using load cell, which reflects the health condition of the plant under supervision with time. Supervised learning algorithm is invoked for analyzing the data obtained using the hardware set-up for decision making, based on the pre-processing of data obtained from well-known database. The entire set-up is extremely beneficial for farmers for real-time assessment of agricultural condition, which is directly correlated with plant growth.

Keywords Real-time monitoring · Agricultural condition · Sensor network · Supervised learning · Production improvement

---

A. Deyasi (✉)
Department of Electronics and Communication Engineering, RCC Institute of Information Technology, Kolkata 700015, India
e-mail: deyasi_arpan@yahoo.co.in

S. Dhar · H. Roy
Department of Information Technology, RCC Institute of Information Technology, Kolkata 700015, India

P. Saha
Department of Plant Pathology, Bidhan Chandra Krishi Viswavidyalaya, Bardhaman 713101, India
e-mail: saha.poly@bckv.edu.in

A. Acharyya et al. (eds.), *Real-World Applications and Implementations of IoT*, Studies in Smart Technologies, https://doi.org/10.1007/978-981-97-8627-5_12

1 Introduction

The growth of any plant is subjected to several levels of uncertainty, where environmental conditions play the most significant role in determining the effective outcome. A lot of research works has been carried out over the last decade to predict the influence of various environmental status, precisely in real-time basis, so that growth rate monitoring can be made technology-dependent and therefore, can help the people in taking right decisions at real-time basis. However, practical implementation of most of the work has failed, owing to mostly the vast differences between laboratory conditions and practical agricultural environment [1–3]. Apart from geographical impact, disease challenges and soil degradation issues; several other key factors demand on-field solutions, predicting key environmental thresholds upto a limit of accuracy [4, 5], and characterization of soil condition w.r.t to the crop under consideration.

Implications of sustainable agriculture are the major demand for today's framework of modern agriculture, where contributions to sustainable practices [6–8] by recommending efficient use of natural resources, reducing dependency on chemical inputs, and improving soil health have severe influence on quality of outcome. Moreover, enhancement of food security is critically based on policy recommendations. Location of the field in terms of latitude and longitude makes a clear distinction in terms of crop yield, and that requires identification of parameters that should be monitored using external sensor network, corresponding on-field analysis, and corresponding decision implementation. In this context, a detailed literature review is required for studying the recent involvements invoked.

2 Literature Review

Jurišić et al. described the sensor network in precision farming [9]. It includes the most important and widely utilized agricultural sensors. In addition, the document describes the primary sensor types based on their architecture, the electromagnetic spectrum's recorded range, and the methods for detecting, logging, quantifying, and representing the energy that is observed. Deeper knowledge of remote sensors and their benefits has been made possible by the growth of remote research. The sensors that are mounted on machinery for crop protection, fertilization, and soil testing, as well as machinery for picking crops, have been examined in relation to precision farming. The extensively used sensors OptRx, ISARIA, and VRT technologies are shown in the study.

The paper's findings evaluate the information gathered by sensors and processed to create maps for agro-technical activities. The utilization of maps reduces the need for human resources, improves data collection capabilities, sharpens agricultural practices, and ultimately lowers the cost of finished goods. Over the past ten years, technological advancements have made it possible to produce technology with variable application standards (VRT), which allows input optimization based on current needs.

Ramya et al. [10] described that traditional farming approaches have proved woefully insufficient to meet the growing demands. As a result, implementing cutting-edge technologies is now essential to improving agricultural standards. In order to measure variables including soil pH, temperature, pressure, light intensity, and moisture content, this research suggests a smart farming system in a small, contained space. Several sensors are placed in strategic locations. This is designed so that anyone could set it up for the least amount of money. As a result, productivity increases as long as the plants have the ideal environmental conditions.

Rajak et al. [11] showed that modern sensors and the Internet of Things (IoT) have the potential to increase agricultural output and reduce financial loss. Research from all around the world has adequately shown how integrated IoT-smart sensors may be used to monitor temperature, humidity, moisture content, and other environmental parameters that are vital to crop growth. Utilizing automated sensors, greenhouse gasses like carbon dioxide and methane are also detected. Smart farming also makes it possible to measure the quantity of nitrogen present in the soil, which aids farmers in figuring out how much fertilizer to apply to their fields. Unmanned aerial vehicles and some IoT-enabled equipment are helpful in providing precise surveillance of insect attacks and related diseases in agricultural vegetation.

Continuous monitoring and tracking are carried out in recent pat [12] which is based on a model developed from a dataset to recommend the most appropriate crop based on sensor outputs while tracking plant and soil vitals in real time. Farmers and producers will be able to increase output and decrease resource waste thanks to IoT-enabled smart farming. The suggested system is a more dependable idea that is simpler to put into practice. It is made up of sensors that can regularly gather critical environmental data, such as temperature, humidity, soil moisture, and soil nutrients. This data is then displayed on an intuitive interface so that farmers and growers can determine the ideal growing conditions for their plants.

Dhanaraju et al. [13] showed that using a recommendation system that is based on a model, developed from a dataset to recommend the most appropriate crop based on sensor outputs while tracking plant and soil vitals in real time. Farmers and producers will be able to increase output and decrease resource waste thanks to IoT-enabled smart farming. The suggested system is a more dependable idea that is simpler to put into practice. It is made up of sensors that can regularly gather critical environmental data, such as temperature, humidity, soil moisture, and soil nutrients. This data is then displayed on an intuitive interface so that farmers and growers can determine the ideal growing conditions for their plants.

Joshi and his group [14] proposed a system that can alert farmers by sending out messages on several channels. With the use of real-time data on temperature, humidity, soil moisture, UV index, and infrared radiation from farms, the solution will help farmers make informed decisions that will boost crop yields and conserve resources (fertilizers and water). ESP32s Node MCU, breadboard, DHT11 Temperature and Humidity Sensor, Soil Moisture Sensor, SI1145 Digital UV Index/IR/Visible

Light Sensor, jumper wires, LEDs, and a live data stream that can be viewed on a serial monitor and Blynk mobile are all used in the device that is being suggested in this paper. This will enable farmers to manage their crop in line with modern farming practices.

Referred to as automated irrigation systems, frost protection systems, remote monitoring systems, decision support tools, and fertilization systems, Gowda and his co-workers [15] proposed IoT based surveillance network for agricultural product monitoring. Because of the information mentioned above, farmers and researchers need to be equipped with a thorough understanding of IoT applications in agriculture. The goal of this project is to improve agriculture through the application of Internet of Things technology and methods. By outlining the need for these tools and describing how they help, this article aims to provide an overview of IoT-based applications in agriculture.

Verma et al. [16] utilized a variety of sensors, as well as wireless technologies such as the Internet of Things (IoT) and an Android application, are used to give farmers a smart platform that allows them to read data from sensors using a microcontroller and receive instructions via a graphical user interface. As a result, we can make use of an IoT platform that provides the main item for internet-based access wherever. The wireless sensor network is made up of infrared and temperature, humidity, and moisture sensors. Utilizing a GSM module and Bluetooth for sensor clustering, additional microcontroller devices are linked to a solar panel or batteries for power delivery.

3 Objective of Present Work

For analyzing any agricultural problem, it is always essential to monitor the soil conditions and environmental parameters during harvesting. Soil type, pH levels, nutrient content, and soil moisture are required to be monitored at real-time basis, whereas temperature, precipitation, humidity, and sunlight conditions are key in that particular geographical territory based on the type of crop to be produced. The present work is solely concentrated on potato harvesting, and therefore, detailed statistical analysis and corresponding adaptation strategies for optimum growth rate prediction. The novelty of our proposed solution essentially depends on the previous findings, as depicted in various literature so far. The following sections are therefore chronologically arranged with an impact analysis of external factors at the beginning, followed by recent works and surveys as found in trusted sources, and henceforth, summarized with the deviations and shortcomings which is concluded by our own approach.

The present solution consists of an ESP32-based IoT enabled device with an integrated PH sensor, TDS probe for water quality, room temperature and humidity sensor, and load cell. Although the proposal may not be perfectly accurate, but owing to ease of implementation, portability during on-field requirement, and unlimited potential for future customization with an approximate cost of Rs. 24,000/- makes it as a customized independent device. With the availability of cloud storage, the real-time data can be stored for further analysis, and also be displayed in OLED screen.

4 System Architecture

The architecture can be separated into several layers such as the application layer which is the layer below the ESP32. The composite architecture is represented in Fig. 1. The user starts with turning on the device and after that the device starts by default the button press is set to 1. The button presses 1 shows the room temperature of the environment. Next with every button press the counter will increment one value and will cycle through 1–6 and then again to 1. From the user's perspective with each button press the selected sensor will change and that sensor's value is going to be displayed in the OLED display.

The 1st and 2nd button press the use will select the DHT11 sensor which will show as mentioned previously will 1st display the room temperature with graph and 2nd button press will exhibit humidity of the environment. The 3rd button press will select the resistive soil moisture sensor and this will show the soil moisture content.

The 4th button press will select the PH sensor which will select the soil and its water's PH level. The 5th button press will select the water TDS sensor where TDS

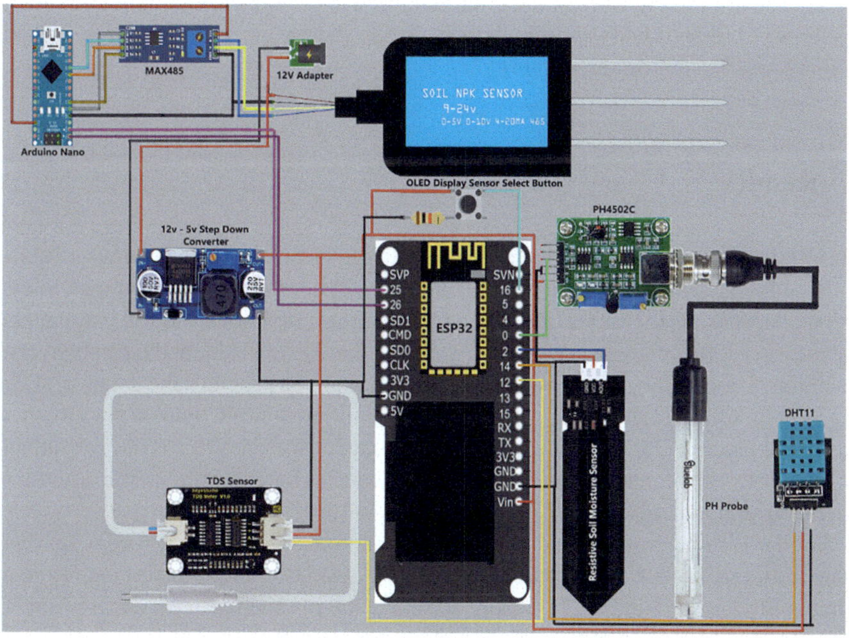

Fig. 1 System architecture

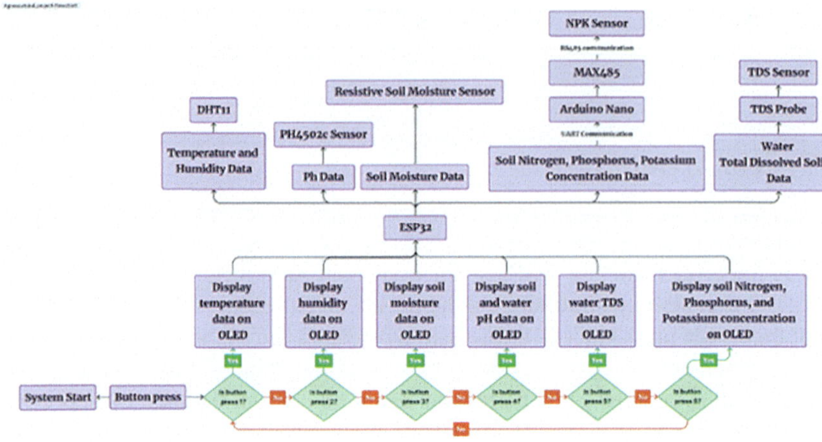

Fig. 2 System workflow

stands for total dissolved solids. The last button is correlated with NPK sensor which will show the soil N, P and K values where N = nitrogen, P = Phosphorus, and K = potassium.

All the sensor after selection will send the data to ESP32 through a single analog data wire except the NPK sensor and DHT11 sensor where how the NPK sensor sends data will be discussed in the circuits section and other than that DHT11 uses a single-wire bidirectional (semi-duplex) protocol. The overall workflow, as described in this section, is represented in Fig. 2.

5 Results

The present architecture provides real-time data for the potato during various seasons in the same geographical location. More precisely, data is taken during summer, rainy, and winter seasons for harvesting, and all the external conditions and soil parameters are measured. Results are graphically represented in Figs. 3, 4, 5, and 6 respectively.

Figure 3 shows the data for summer season. Figure 3a provides the data for TDS, Fig. 3b for soil pH and water pH, Fig. 3c for soil moisture. Compared with other seasons, average TDS values are quite high in summer seasons. This is due to gradual decrease of water content in the soil. Soil pH, water pH and soil moisture variations are noted simultaneously.

Similar results are exhibited in Fig. 4a, b, and c respectively for winter season. Due to slightly higher water content in soil, TDS level decreases, as revealed from a comparative study with Fig. 3a. Figure 4b shows soil pH and water pH during winter, whereas Fig. 4c depicts soil moisture under same condition.

Similarly, data for rainy season are shown in Fig. 5a, b, and c respectively.

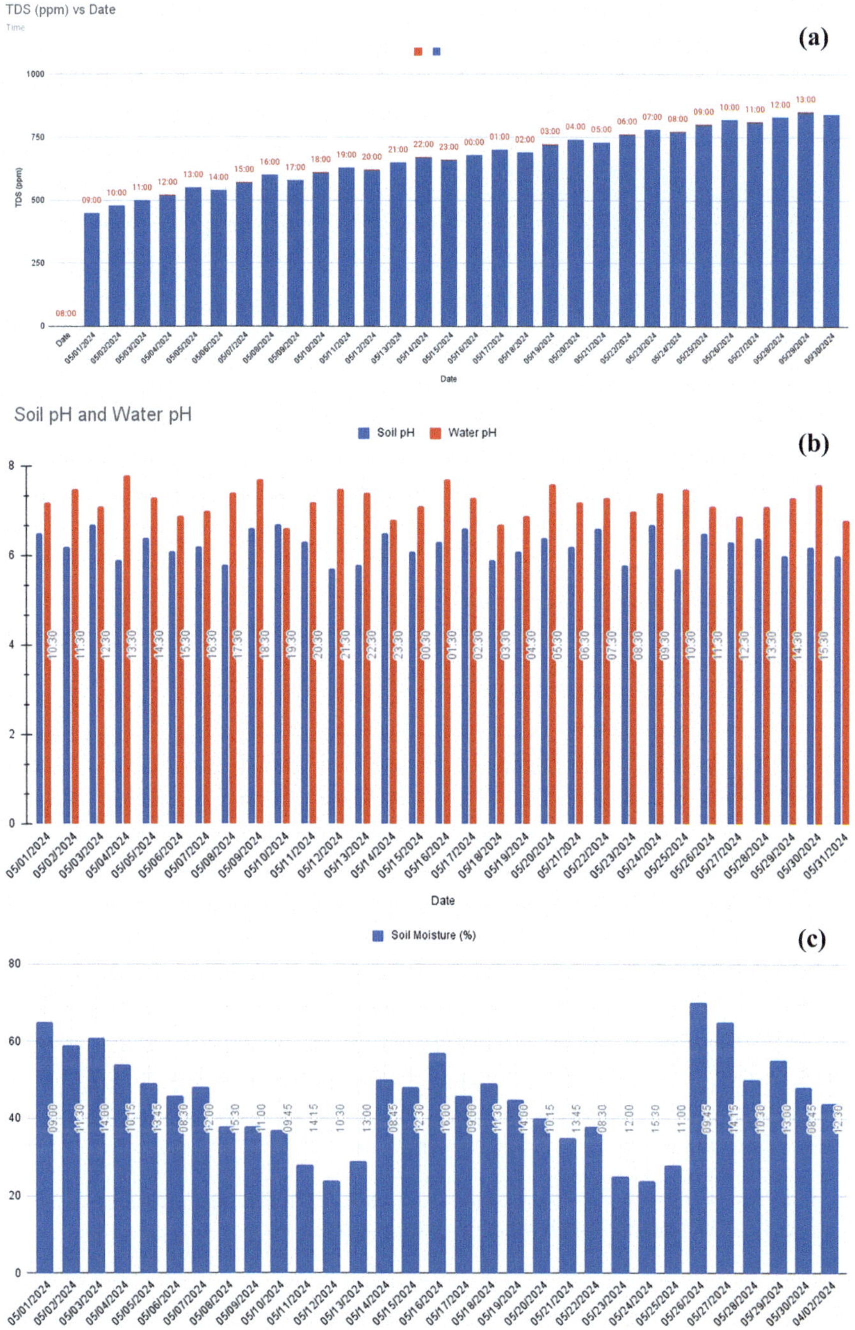

Fig. 3 **a** Change of TDS level during summer season. **b** Soil pH and water pH measured during summer season. **c** Soil moisture measured during summer season

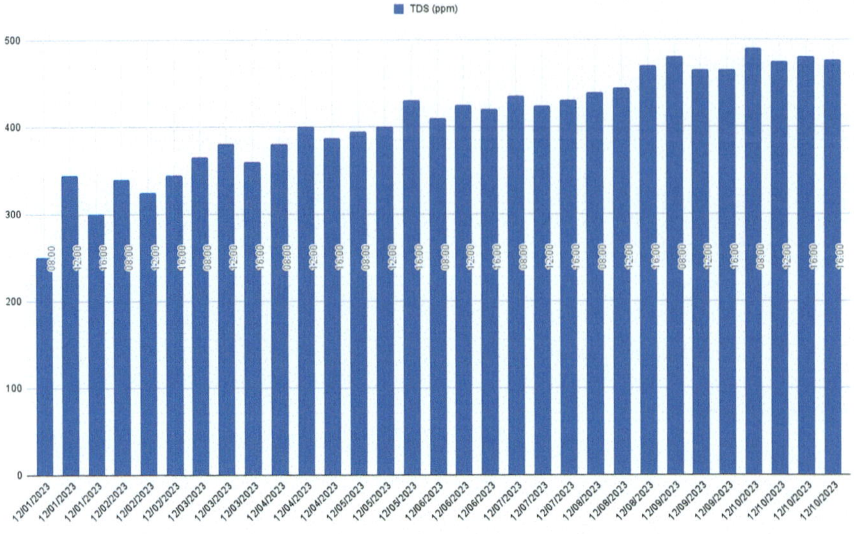

(a)

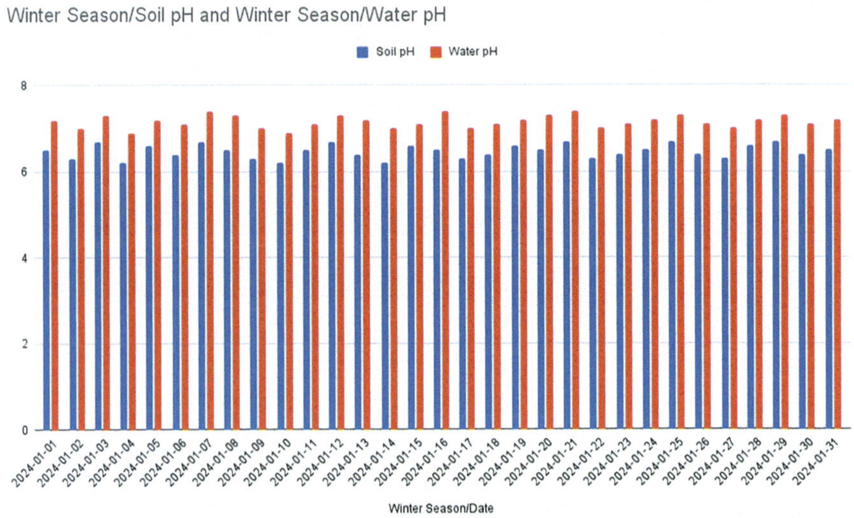

(b)

Fig. 4 **a** Change of TDS level during winter season. **b** Soil pH and water pH measured during winter season. **c** Soil moisture measured during winter season

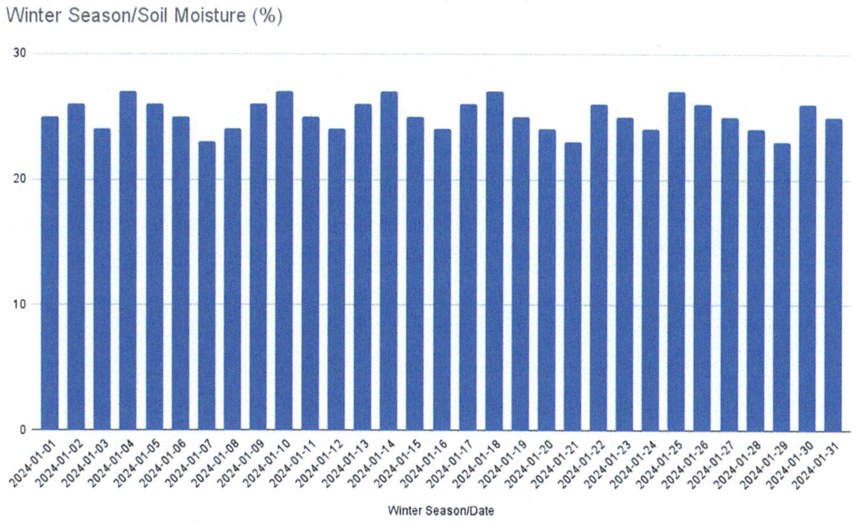

(c)

Fig. 4 (continued)

Figure 6 shows the variation of NPK levels. Observations in the detailed study reveals that

i. Nitrogen levels are consistently high, around 175 units each day. This stable supply is crucial for potato growth, supporting vegetative and reproductive development.
ii. Potassium levels remain steady at 100 units throughout the period. Potassium is important for the overall health of crop, aiding in water regulation and disease resistance.
iii. Phosphorus levels are significantly lower than Nitrogen and Potassium, at about 10 units daily. Despite being lower, this nutrient is essential for root development and energy transfer within the plants.

6 Conclusion

In conclusion, the analysis of environmental parameters and soil conditions for gram harvesting is a crucial endeavor with far-reaching implications for food security, economic development, and environmental sustainability. By unraveling the complexities of chickpea cultivation, this project aspires to empower farmers and stakeholders with knowledge that fosters resilience and prosperity in the agricultural sector. The project will employ a multidisciplinary approach, integrating field studies, soil analysis, and data from meteorological stations. Researchers will collaborate with farmers to collect real-time data throughout the chickpea cultivation cycle,

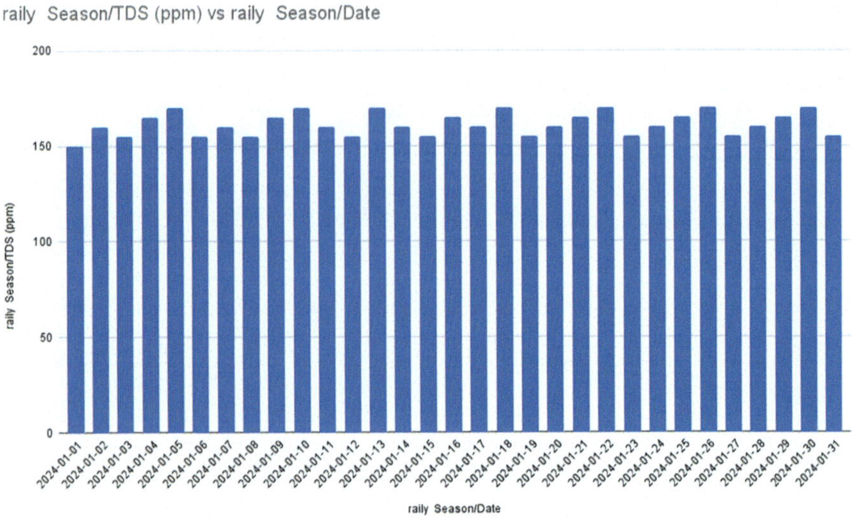

(a)

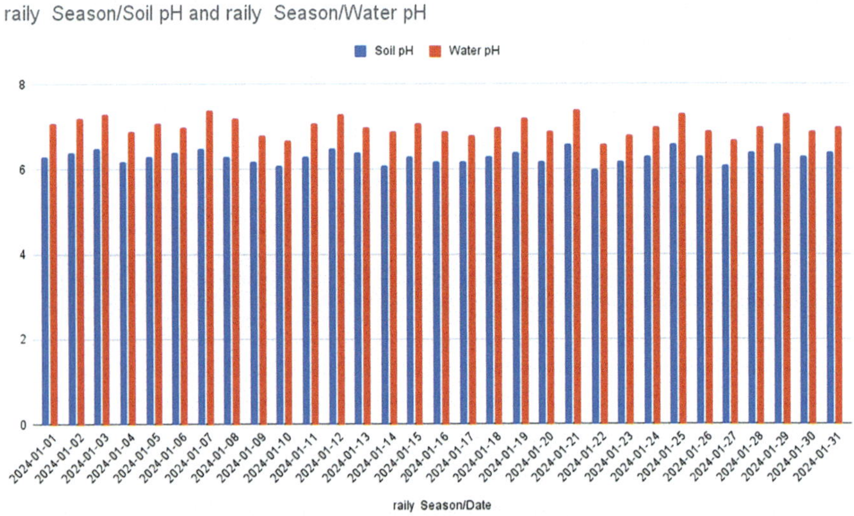

(b)

Fig. 5 **a** Change of TDS level during rainy season. **b** Soil pH and water pH measured during rainy season. **c** Soil moisture measured during rainy season

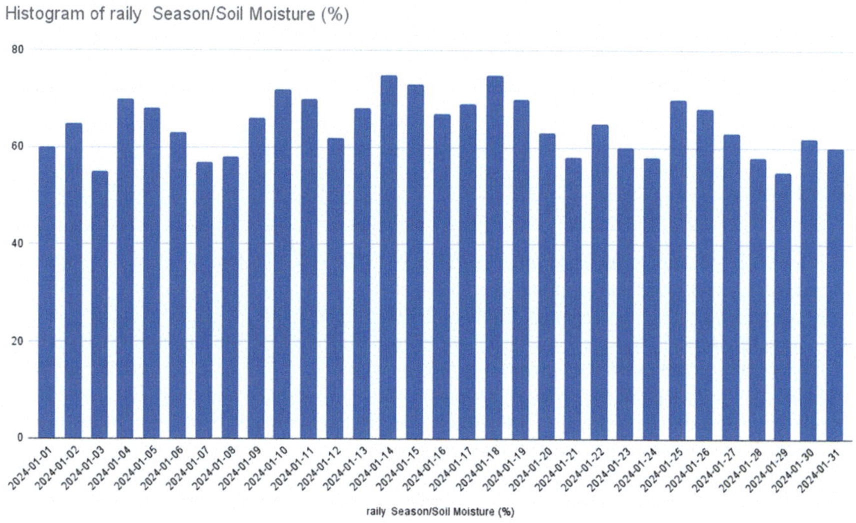

(c)

Fig. 5 (continued)

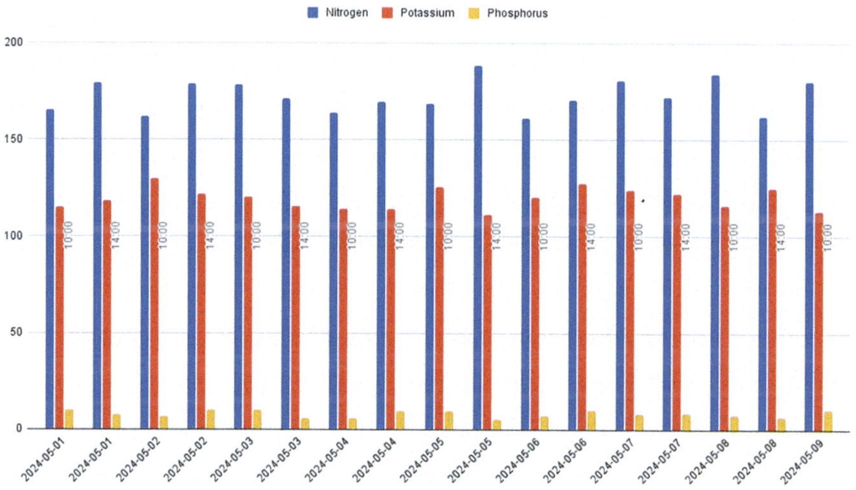

Fig. 6 NPK variations for a consecutive 9 days reading

considering variations in environmental and soil conditions. Statistical analyses and modeling techniques will be applied to derive meaningful patterns and correlations. The outcomes of this project aim to provide farmers with actionable insights into optimizing chickpea cultivation. This includes the development of guidelines for planting, irrigation, and fertilization based on local environmental and soil conditions.

Acknowledgements The work is funded under the grant-in-aid scheme of West Bengal DST-BT vide memo no. 1407(sanc)/STBT-11012(11)/45/2021-ST SEC.

References

1. Ram VVH, Vishal H, Dhanalakshmi S, Vidya PM (2015) Regulation of water in agriculture field using internet of things. In: Proceedings of the IEEE technological innovation in ICT for agriculture and rural development, Chennai, India, pp 112–115
2. Farooq MS, Riaz S, Abid A, Umer T, Zikria YB (2020) Role of IoT technology in agriculture: a systematic literature review. Electronics 9(2):319
3. Saha A, Bhattacharyya S, Ghosh S, Deyasi A (2019) Environmental condition based auto-adjusted robot design for agricultural purpose. In: 3rd international conference on electronics, materials engineering and nano-technology
4. Muangprathub J, Boonnam N, Kajornkasirat S, Lekbangpong N, Wanichsombat A, Nillaor P (2019) IoT and agriculture data analysis for smart farm. Comput Electron Agric 156:467–474
5. Foughali K, Fathallah K, Frihida A (2018) Using cloud IoT for disease prevention in precision agriculture. Procedia Comput Sci 130:575–582
6. Zeweld W, Huylenbroeck GV, Tesfay G, Azadi H, Speelman S (2020) Sustainable agricultural practices, environmental risk mitigation and livelihood improvements: empirical evidence from Northern Ethiopia. Land Use Policy 95:103799
7. Siebrecht N (2020) Sustainable agriculture and its implementation gap—overcoming obstacles to implementation. Sustainability 12(9):3853
8. Muhie SH (2022) Novel approaches and practices to sustainable agriculture. J Agric Food Res 10:100446
9. Jurišić M, Plaščak I, Željko B, Radocaj D (2021) Sensors and their application in precision agriculture. Tehnički glasnik 15(4):529–533
10. Ramya R, Sandhya C, Shwetha R (2017) Smart farming systems using sensors. In: IEEE technological innovations in ICT for agriculture and rural development, Chennai, India
11. Rajak P, Ganguly A, Adhikary S, Bhattacharya S (2023) Internet of things and smart sensors in agriculture: scopes and challenges. J Agric Food Res 14:100776
12. Cheruvu B, Latha SB, Nikhil M, Mahajan H, Prashanth K (2023) Smart farming system using NPK sensor. In: 9th international conference on advanced computing and communication systems, Coimbatore, India
13. Dhanaraju M, Chenniappan P, Ramalingam K, Pazhanivelan S, Kaliaperumal R (2022) Smart farming: internet of things (IoT)-based sustainable agriculture. Agriculture 12(10):1745
14. Doshi J, Patel T, Bharti SK (2019) Smart farming using IoT, a solution for optimally monitoring farming conditions. Procedia Comput Sci 160:746–751
15. Gowda D, Prabhu S, Ramesha M, Kudari JM, Samal A (2021) Smart agriculture and smart farming using IoT technology. J Phys Conf Ser 2089:012038
16. Verma S, Kumar M, Rajesh R (2020) Vincent, smart-farming using internet of things. J Comput Theor Nanosci 17(1):172–176

Chapter 13
IoT-Driven Advanced Lighting Detection and Fire Response System for Forest and Wildlife Protection

Md Tarikul Islam Sheam and Erin Tabassum Arpa

Abstract Among all other common catastrophes in recent times, forest fires have emerged as a significant issue. Natural calamities capable of igniting wildfires include lightning strikes, volcanic eruptions, sparks from rock falls, etc. Lightning is a common cause of wildfires. Forest fires are responsible for 20% of global carbon emissions each year. This is equal to the carbon dioxide emitted by every car on the world's roads. In the past, sustaining life in the forest relied on manual methods that often caused serious damage to plants and trees. Fast, efficient, and accurate detection is essential in solving these problems. The purpose of this chapter is to introduce the use of Internet of Things (IoT) technology for early detection of electrical fires in the forest to reduce the spread of fire. This scheme points toward upgrading woodland security by consequently recognizing lightning inside a particular area and observing for fires within, During any occurrence, the framework instantly informs the specialists and quickly empowers successful reaction measures. In this research, we are going utilize the complete architecture of modules and circuits that will detect lightning frequency and give output to the system, then the system will investigate its range, and after that, it will immediately convey the information to the authority. The implementation of the output of this research will give a new dimension to the forest protection system to keep the ecosystem safe more efficiently.

Keywords Internet of Things (IoT) · Forest · Lightning · Frequency · Architecture

Md T. I. Sheam (✉) · E. T. Arpa
Department of Electrical and Electronic Engineering, Faridpur Engineering College, Dr Kaji Motaher Hossain Rd, Faridpur, Bangladesh
e-mail: tarikulislam.sheam@gmail.com

E. T. Arpa
e-mail: eerintabassumarpa@gmail.com

A. Acharyya et al. (eds.), *Real-World Applications and Implementations of IoT*, Studies in Smart Technologies, https://doi.org/10.1007/978-981-97-8627-5_13

1 Introduction

Forest ecosystems, characterized by their lush green canopies and diverse flora and fauna, are not only a source of awe-inspiring beauty but also play a pivotal role in maintaining ecological balance on our planet. However, amid the tranquillity of these wooded landscapes, lurks a powerful force of nature—lightning. Lightning, a natural phenomenon born from electrostatic discharges in the atmosphere, possesses the potential to wreak havoc upon forests, igniting flames that can transform the serene tranquillity of the wilderness into scenes of chaos and destruction in an instant. In this introduction, we delve into the intricate relationship between forests, lightning, and fire, exploring the profound impact of these natural phenomena on ecosystems, wildlife, and human communities worldwide. Lightning is a phenomenon that occurs because of the rapid discharge of a large electrical spark in the air or between the air and the ground when the difference is very large. This discharge typically involves an initial current flow of millions of amps, followed by a train of pulses of decreasing energy. With lightning carrying up to 100 million volts and reaching temperatures of 50,000 °F, its effects on both humans and the environment are profound [1]. For humans, a direct strike can be catastrophic, resulting in internal burns, organ damage, explosions of flesh and bone, and nervous system damage. In forested areas, lightning can have similarly devastating effects. When lightning strikes trees, the extreme temperatures can boil water and sap within the tree, generating steam and causing bark and wood fibers to explode. If the outer layer of bark is soaked from heavy rainfall, the lightning may travel along the outside of the tree to the ground, resulting in minimal damage. However, intense lightning bolts can split trees in two and cause them to burst into flame from the inside out, leading to forest fires that can ravage vast expanses sof woodland [2]. Thus, the interaction between forests, lightning, and fire is a complex and often destructive one, highlighting the delicate balance between the forces of nature and the resilience of the ecosystems they shape. Understanding this relationship is crucial for mitigating the impacts of wildfires on forests and the communities that depend on them for their livelihoods and well-being.

In addition to the devastating effects lightning can have on forests, there's a growing concern regarding the detection of lightning strikes and the subsequent risk of forest fires they pose. Detecting lightning strikes in real-time and promptly alerting authorities and forest management teams is crucial for mitigating the potential impact of resulting fires. Furthermore, there's an ongoing exploration and discussion about integrating new and revised technologies, including those integrated with the Internet of Things (IoT), to enhance our capabilities further in lightning detection and wildfire prevention. We will now delve into the exploration and discussion surrounding these advancements.

2 Occurrence of Lightning and Characteristics

Lightning strike is one of the oldest events. This is the glow seen in the sky caused by a sudden discharge of electricity into the atmosphere during a storm. The production environment is the cloud. Lightning can be seen as a steady flow of electricity in the air along high-speed lines, high-speed lines, and airborne plasma lines. Due to the separation of positive and negative, high voltage is created in space. Therefore, the main source of lightning is storm clouds, also known as cumulonimbus clouds. The formation and payment of thunderstorms begin with the formation of cumulus clouds, which are formed by the rise of warm air. When these clouds grow vertically, they produce rain, including ice, hail, and rain. Collisions of particles in clouds cause charges to separate; lighter ice becomes positive, and heavier ice becomes negative. This gives a positive result to the upper clouds and a negative result to the middle and lower clouds. Additionally, a small amount of positive charge develops near the base of the cloud. Charge imbalance between cloud and ground causes lightning [3, 4].

The lightning dischargement can be produced within the cloud, or between a cloud and the ground [2]. Lightning from thunderstorms is induced in a strong electric field which can stay completely within the cloud when the charge regions are similar in strength (balanced) or can reach the ground when one of the regions is much stronger than the other (unbalanced) [2].

Lightning forms in clouds through various types of discharges, (i) Intracloud-lightning intracloud discharge, (ii) Cloud-to-air lightning discharge (iii) Cloud-to-ground lightning discharge.

Cloud-to-ground lightning can start from a negative conductor or a positive charge, cloud-to-ground lightning can lead to a lower charge and is divided into four types according to the polarity of the charge sent to the ground and the direction of the initial lightning. Leader. These types are (a) ascending lightning, (b) ascending negative lightning, (c) descending lightning, and (d) descending lightning. By depositing funds through a special method, the administrator creates a way to replace the price in the cloud and on the ground [4]. Upward lightning also occurs from tall objects within the cloud charge regions. In this chapter, we are going to thoroughly discuss about cloud-to-ground lightning as it's the only type to contact with the ground and our aim is to detect lightning-caused fire using sensor network and IoT (Fig. 1).

3 Analyzing Large Wildfires (A Statistical Study of Cases in the US)

The current national statistics for wildfires in the US reveal a challenging situation. With 7 reported incidents, there's a total of 31 large fires across the country. The emergence of new large fires is particularly concerning, with a staggering 9,458 incidents recorded recently. Year-to-date data reflects a significant wildfire burden, with 21,352 incidents reported thus far. These fires have ravaged approximately

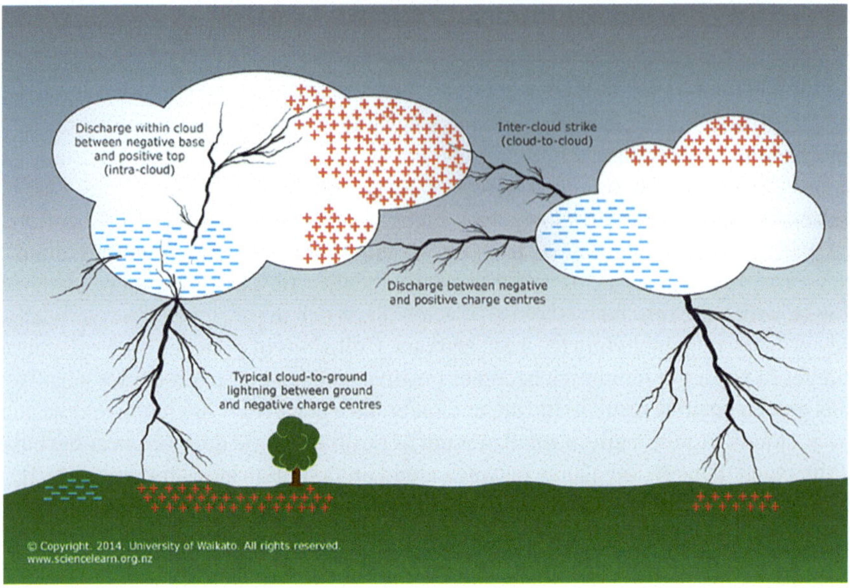

Fig. 1 Illustration of lightning generation

21,352 acres of land, with large fires alone accounting for 477 acres burned. To combat these wildfires, a massive workforce comprising 1,754,439 personnel has been mobilized. Despite these efforts, the year-to-date acres burned stands at an alarming 1,754,439 acres, underscoring the magnitude of the challenge faced by firefighting agencies and communities across the nation [5] (Fig. 2).

4 Challenges and Limitations

The necessity of forest fire management is unavoidable because uncontrolled wildfires can have severe negative impacts on the climate, individual health, livelihoods, assets, air and water quality, and biodiversity. It is tracked that 350 million hectares of land were affected by fire in 2000, leading to a wide range of damage to civilization and the environment. Therefore, detection of forest fires is a very important problem in the pre-fighting process. Integrated fire management (IFM) involves five key elements: review, risk reduction, readiness, response, and recovery, with forest managers playing crucial roles in each aspect [8] (Fig. 3).

The need for financial, institutional, and technological resources, capacity development, and transboundary cooperation are some challenges in forest fire management. Some technical equipment that is lacking in most forest fire management system [8] include:

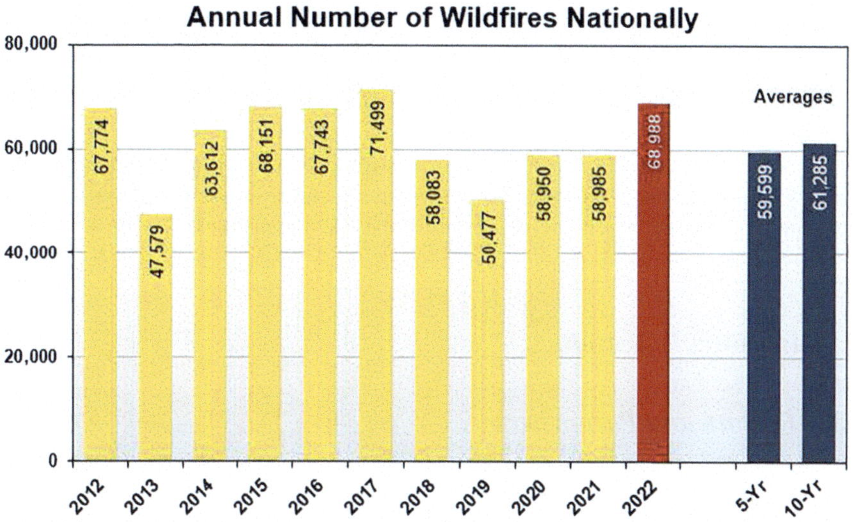

Fig. 2 The number of U. S. wildfires in 2022 (red bar) was higher than the 5- and 10-year averages (blue bars on right) [6]

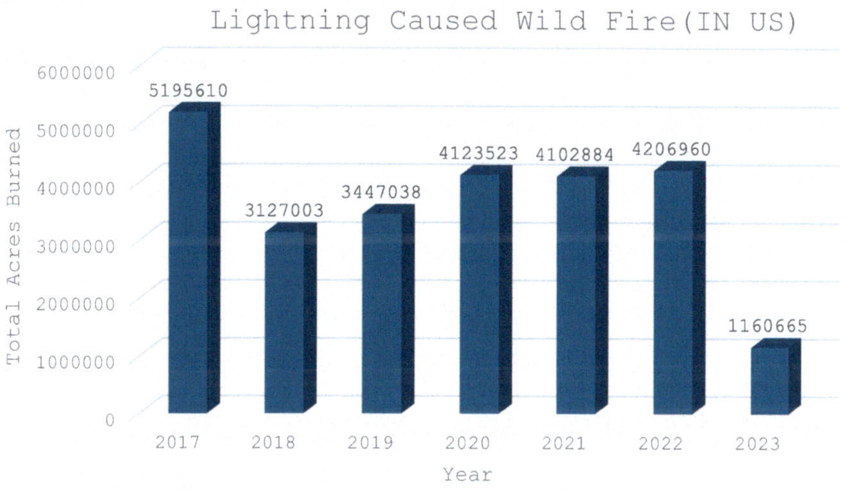

Fig. 3 An overview of the annual variation in the total extent of wildfire damage caused by lightning across the United States [7]

1. Advanced fire detection system.
2. Robust Communication.
3. Remote Forest behavior mapping tool.
4. Sensing and mapping tool.

5. Lack of adequate resources for effective implementation of policies and legal arrangements related to forest fire management.

Due to this lack in current technology, it is mandatory to invest in research and development to improve the immediateness and performance of forest fire monitoring systems. Furthermore, collaboration with experts in both technology and forestry management is essential to ensure the accurate implementation of legal and constitutional arrangements related to forest fire management.

5 Lightning Detection

5.1 Factors and Parameters

The main three factors [9] used in latest technologies are-

1. *Change in electromagnetic field*: Electric field meters (EFM) are devices that measure changes in the electric field of the Earth's atmosphere. These changes in the electric field occur in response to the combination of charges in the storm cloud and the onset of lightning. EFM detects and reports changes in electrical charge from a predetermined point to the lightning value, indicating the presence of a lightning strike. While EFMs provide important information about weather conditions for lightning, they can have limitations such as narrow alerts and can interfere with false alarms from other sources such as dust storms.
2. *Optical Radiation*: Optical monitors are specialized instruments that detect the light flashes produced by lightning discharges, particularly those occurring within clouds (cloud-to-cloud lightning). Cloud-to-cloud lightning often precedes cloud-to-ground lightning strikes, making optical monitors valuable tools for early detection of developing thunderstorms. Optical monitors are effective in providing visual confirmation of lightning activity, which can be useful for meteorologists and weather forecasters in tracking storm development and severity.
3. *Electromagnetic frequency emissions*: RF detectors are designed to measure the radio frequency emissions generated by lightning discharges. These emissions occur as a result of the rapid acceleration of charged particles during a lightning strike. By analyzing these RF emissions, the detector can detect the occurrence of lightning RF detectors can also calculate the distance and direction of the lightning based on the location of the detector. RF detectors are particularly useful for providing early warning of approaching thunderstorms and lightning activity, allowing for timely safety precautions to be taken.

5.2 *State of the Art*

Modern lightning-detecting technology contains a great number of specific devices to help in detecting and monitoring lightning. The RF sensors monitor electric charge from past lightning strikes and locate the source by measuring the varying frequency and direction. These detectors are irreplaceable tools for forecasting approaching thunderstorms and lighting danger. Instrument the invaluable multi-station inference meters which are professional in research operations. They supply granular information regarding the lighting manifestation enabling scientific studies to be carried out and therefore advancing atmospheric processes knowledge. Employed in various ground-based RF sensors forming a wide network measure the lightning. They encompass large spatial ranges like continents or whole countries for accurate tracking of lightning occurrence over long distances. In this case, electric field meters (EFMs) recognize potential changes in the earth's electricity and operate ahead of lightning sensors. They report variations of the prepormed threshold levels which point to the presence of the lightning coaching power value. EFMs, as useful as they are in providing information about lightning-prone conditions, do sometimes have limited reporting range or false alarms caused by a dust storm. On the other hand, optical monitors are especially good at detecting light traveling from one cloud to another, which usually foreshadows cloud-to-ground strikes. This front-end early detection ability greatly assists the task of meteorologists in tracking the development and severity of the storm. Combo approaches merge the technologies under one roof to exploit the strengths of each to its advantage. Such systems therefore demonstrate increased lightning detection skills, supplying comprehensive coverage and greater accuracy. Further along, weather information services provide the subscribers with lightning data that they have received from the network stations on their services. They complement important real-time information systems of operational weather meteorologists and weather forecasters through the application of which meteorologists are able to enhance the prediction and management of severe weather events. Overall, combining these diverse lightning detection technologies and services plays a crucial role in safeguarding lives and property from the impacts of lightning-related hazards (Fig. 4).

6 Potential of IoT Integration

We discussed about the detection system of lightning, we also discussed how lightning can cause wildfire. The combination of lightning and wildfire detection using IoT is going to be the next generation technology. Integration of the Internet of Things (IoT) technology in the detection and management of forest fires caused by lightning is an innovation that can bring forth certain key benefits that might revolutionize whole wildfire combustion prevention, reading, and response strategy.

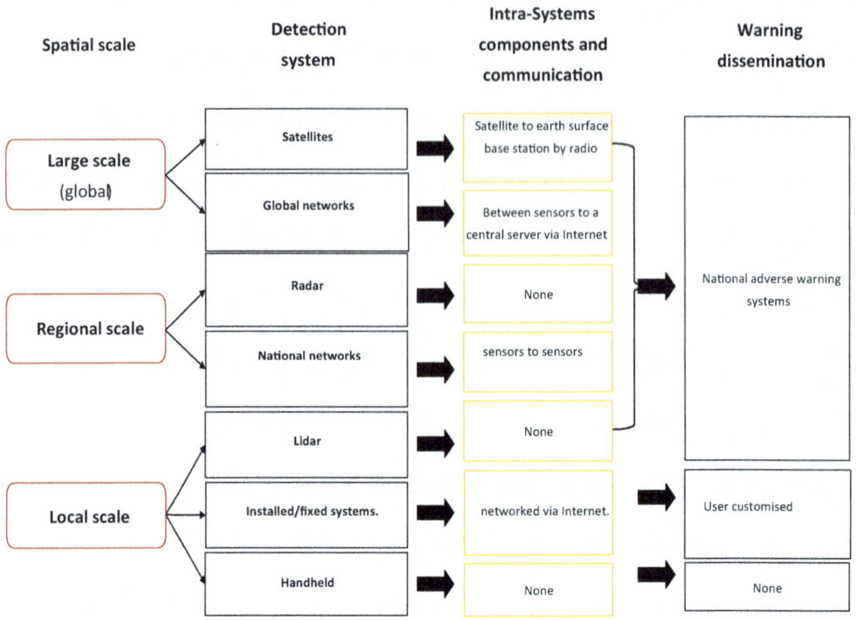

Fig. 4 The spatial characteristics of existing lightning detection systems and their associated warning and alert dissemination capabilities [9]

Early Detection and Rapid Response: IoT sensors with lightning detection capabilities will be able to see and give directions to the lightning strikes in the forest. Such sensors can identify the electromagnetic radio signals that are assigned to thunder and then relay the information to the centralized cloud computer immediately. Partly due to the quick determination of lightning strikes, firefighting agencies have an opportunity to quickly deploy resources, assign responsibilities, and respond upfront to potential wildfire ignitions, thus preventing transmission and mitigating damage.

Remote Monitoring in High-Risk Areas: IoT sensors are able to be installed in forests that are hard to reach or inaccessible and are high-risk areas for lightning strikes or may be installed in the forests which are far away or inaccessible and are prone to lightning strikes. These sensors will be able to continuously monitor environmental conditions. Authorities use these databases to find out the areas that may be affected by the lightning-induced wildfire. Therefore, they implement various measures such as enhanced patrol and prescribed burns to curb the risk.

Integrated Sensor Networks: The global IoT sensor network can easily cover the whole range of forest areas so that it won't be difficult to build a comprehensive wildfire monitoring system. The forecasting networks could be composed of several types of sensors that include lightning detection devices, weather stations, and thermal imaging ones connected via wireless networks of communication. Authorities can obtain insights into wildfire risk factors by gathering data from sensors installed in many places which can help them to manage fire from the beginning and determine the best firefighting strategies.

Predictive Analytics: By utilizing data analysis algorithms on historical weather patterns, lightning strike data, and environmental conditions used for the development of predictive models for lightning-induced wildfires, we can also create some predictive models. As the machine learning methods are implemented, the models will be able to estimate the probability of wildfire occurrence in given regions and conditions, which depends on the current and future weather conditions. Such a competency to predict enables authorities to allocate resources in advance and deploy preventive measures early to keep the magnitude of the wildfire impact to the minimum.

Real-Time Alerts and Notifications: At IoT-connected systems, it is possible to regulate and produce real-time alarms and notifications, consequently of lightning strike detection and potential wildfire induction. Such alerts would be issued to firefighting engineers, firefighters, and local communities through social apps, emails, SMS, besides other communication channels. Through prompt alerts, authorities do have opportunities to organize the evacuation progress and rescues. This reduces the possibility of fatality and property loss.

Enhanced Resource Allocation: Through the IoT data analysis, firefighting bodies will be able to plan effectively and use their resources in a better way in times of wildfire. Timely data on lightning activity, fire behavior, and environmental conditions helps to understand where authorities can deploy most fire crews and equipment, aircraft, as well as other resources to avoid losses and injuries. The decisions are then made quickly and accurately.

In general, the combination of IoT technologies and lightning-originating forest fires has a potent toolset that increases efficiencies in early detection and monitoring capabilities. Combining the power of IoT sensors, networks, and data analytics, authorities are given an additional factor that will help their ability to detect, respond to, and minimize damage associated with lightning-induced wildfires subsequently, the lives are protected, property is saved, and natural ecosystems are preserved.

7 Architecture of Proposed System

Our proposed system focuses on local-scale detection and alarm mechanisms tailored specifically for fire safety authorities which will be networked via Internet. We broke down the architecture into different units (Fig. 5).

Although the controller, server, and monitoring panel will be using the latest technology, our new novel idea is in the sensing circuit. Our idea is to integrate both the fire and lightning detection together to get more specific information about how the fire ignited (Fig. 6).

The control unit and sensor unit will be combined in the detector module. The control unit will be connected to the internet with its embedded wifi module. The

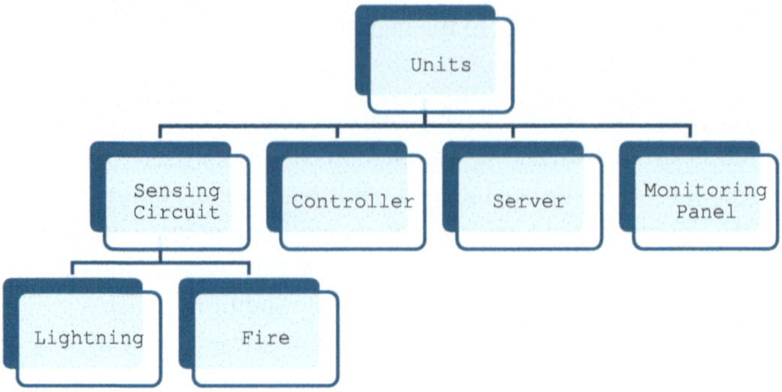

Fig. 5 Units of our proposed system

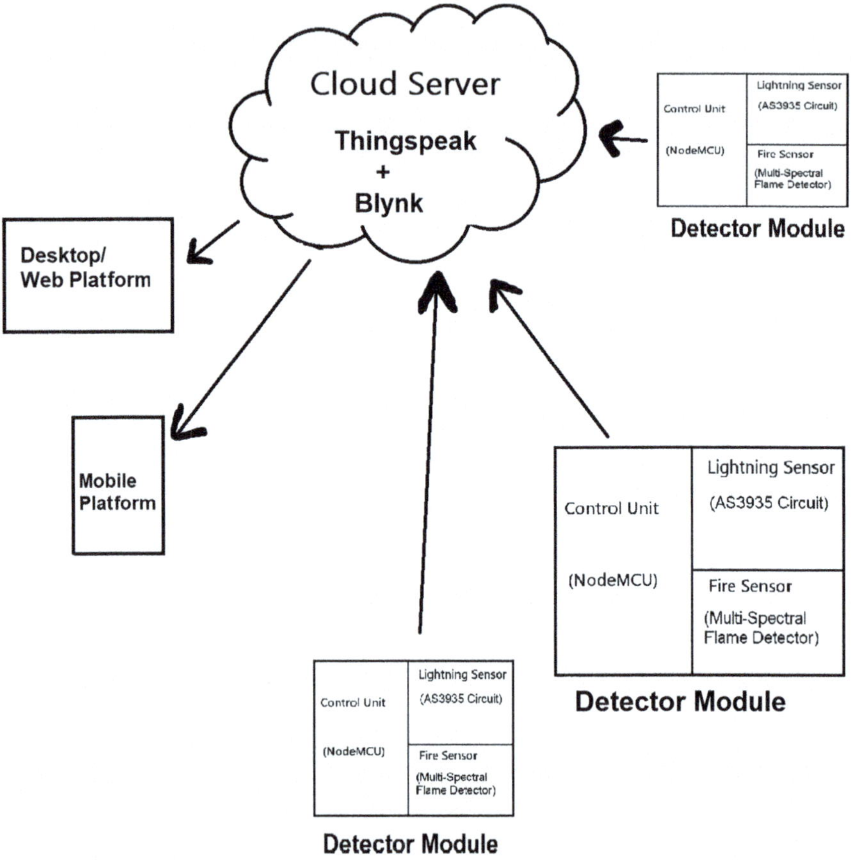

Fig. 6 Architecture for the detection system

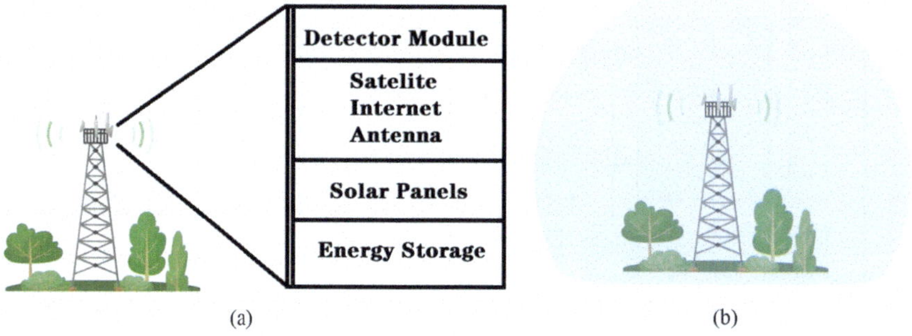

Fig. 7 **a** System components on tower, **b** Tower range sphere

data taken from Fire and Lightning sensor going to be uploaded to the cloud server via Controller. We are using Thingspeak to serve desktop platforms and Blynk for the mobile platform. Both software can visualize and analyze real-time data streams in the cloud. Here we can see that the detector modules will be spread to the jungles or forests. So, these modules will stay in a remote place and inclement weather. We need to place those in a tower to keep those secure and to cover a wide range. The components on the tower for the purpose will be detector modules for collecting field data, a satellite internet antenna to keep the module connected to the internet, solar panels to generate and store energy, and energy storage to supply voltage to the modules and other peripherals. The range of the tower depends on the fire and lightning sensor we will be using. The response time will depend on the components we choose and the speed of satellite internet constellations (e. g., Starlink) (Fig. 7).

7.1 Components

Controller:

NodeMCU (LoLin): It is an Esp8266 SoC-based microcontroller module, which offers better memory and better processor than others at cheaper price. The integrated wifi module makes it an ideal choice for IoT applications. We are using it here because we will need to keep our system connected to the cloud server via Internet.

Key Features:

Its main features include 32-bit architecture, 80 MHz operating clock speed, 3.3 V operating voltage, and 4.5 – 10 V input voltage range. It provides 11 digital I/O pins and 1 digital analog input and supports UART, SPI, and I2C communication protocols. It is suitable for use in various environments between −40 and 125 °C (Fig. 8).

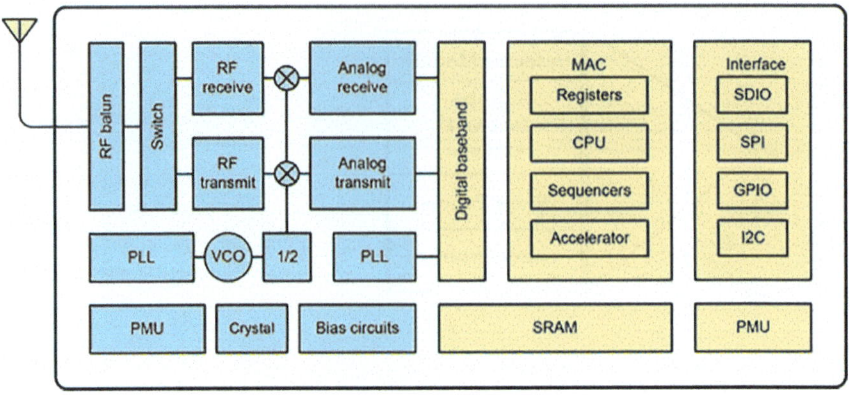

Fig. 8 Functional block diagram of ESP8266

Fig. 9 NodeMCU ESP8266 [10]

Lightning Sensor Circuit:

Lightning Can be detected in many ways, we talked about that in 7.2. Here we will be using EM frequency generated by lightning because of its reliability. We can approach this in two ways: designing our own custom resonance circuit, or utilizing an integrated chip optimized for this purpose. The AS3935 from AustriaMicroSystems (AMS) is the first programmable IC that can detect and analyze airborne lightning and predict the location of lightning during storms [11]. The chip can communicate both in SPI and I2C with the NodeMCU. It also provides noise detection features at a low cost (Fig. 9).

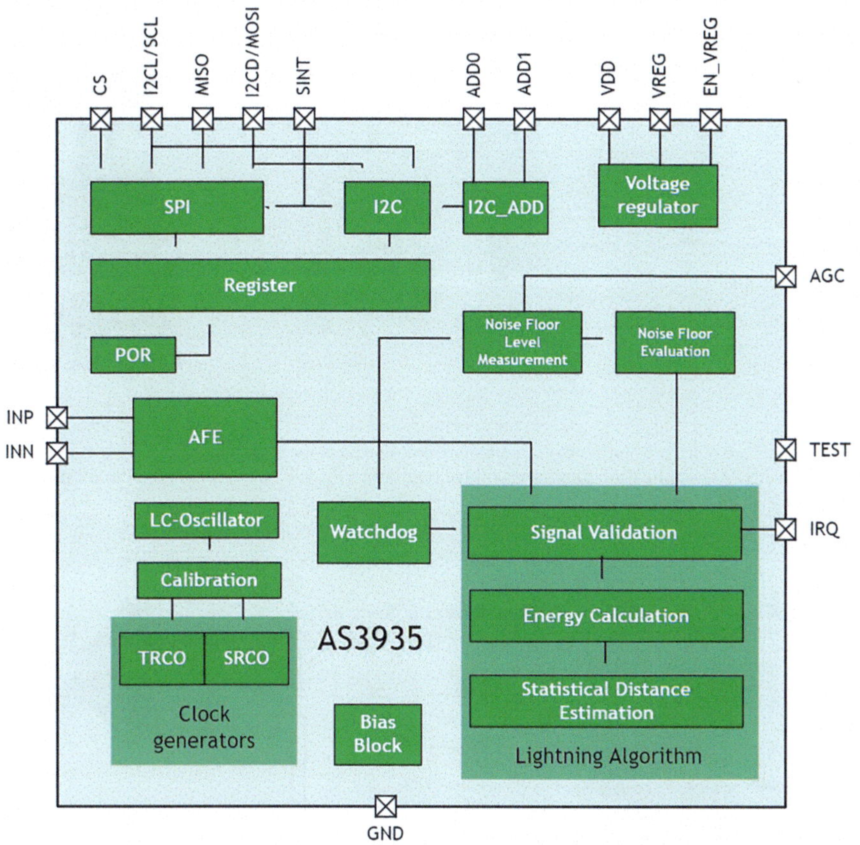

Fig. 10 AS3935 Block diagram [12]

Firstly, the IC enables the detection of both cloud-to-ground and cloud-to-cloud lightning activity within a 40-km range, providing comprehensive coverage. Additionally, its intelligent algorithm includes mechanisms for false disturber rejection, enhancing the accuracy of lightning detection. Moreover, the IC operates within a flexible supply voltage range of 2.4 – 5.5 V, supporting various power modes including power down, listening, and active modes, thereby optimizing energy efficiency. Furthermore, the IC features antenna auto-tuning capabilities, ensuring optimal performance and reliability in diverse environmental conditions. Based on these advantageous features, the IC emerges as the preferred choice for integration into our lightning detection system (Fig. 10).

In this study, we partially present a cost-effective lightning detection system designed to detect lightning storms within a 40 KM range from the system's installation location. The recorded data is subsequently uploaded to the cloud, providing authorities with the capability to monitor it conveniently via desktop or mobile devices (Fig. 11).

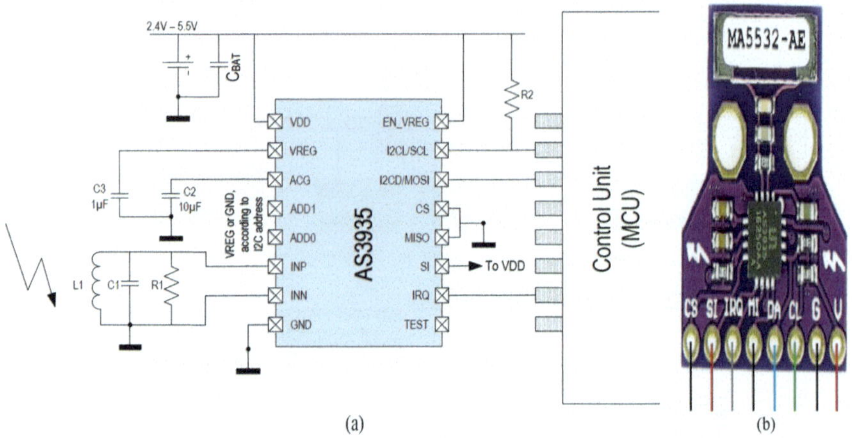

Fig. 11 **a** AS3935 application diagram (voltage regulator ON, I2C active), [13] **b** a circuit module of AS3935

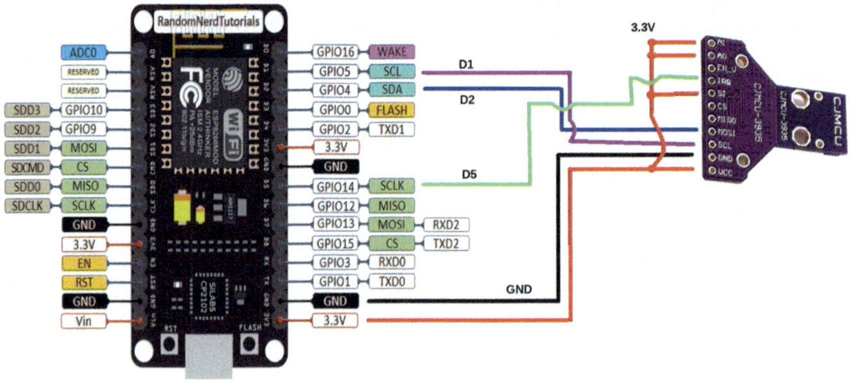

Fig. 12 AS3935 I2C wiring for NodeMCU ESP-12E [14]

In this study, the Inter-Integrated Circuit (I2C) protocol is implemented to facilitate communication between the AS3935 and NodeMCU, utilizing the SCL, SDA, and SCLK signals (Fig. 12).

Fig. 13 Det-Tronics X3301 multispectrum infrared (IR) flame detector

Fire Sensor:

A flame detector is an electronic device developed to identify a flame or fire in a convenient and effective way; It works mainly with infrared radiation. There are different types of Flame sensors available. We will be using Multispectral Flame Detector for this system because of its wide range of coverage. Its going to be both cost-efficient and accurate (Fig. 13).

We will customize the alert system of MFD (Multispectral Flame Detector) and take low voltage (3.3 V) output from the flame detector and input it to the NodeMCU Pin D6.

7.2 Flow Chart

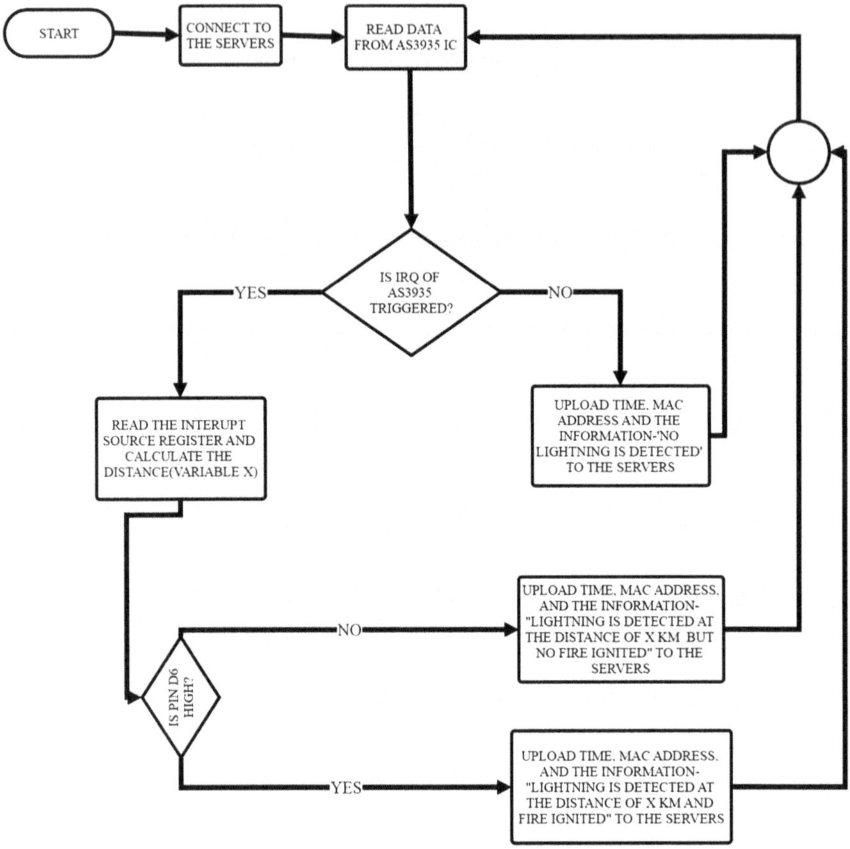

Fig. 14 Flow chart for the program of lightning detection and fire detection over cloud

8 Application of Similar Technology Across Domains

Weaving in lighting detection methods across multiple sectors opens up a huge possibility of having appropriate safety disciplines and reducing lighting lightning strikes associated hazards. With the use of innovative systems of monitoring and automatic notification, we will heighten resilience and security from hazards brought by lightning. Hence, the lives, properties, and infrastructures will be saved.

In case of agricultural enterprises, the system of lightning detection proves to be an effective tool for early warning of possible fire ignition due to lightning strike at fields or warehouses. Such a planned position provides farmers with grounds for preventive measures and reduces the danger of crop impairment and economic losses.

Therefore, the systems of lightning detection can be used for protecting the vital infrastructure among them are power lines, communication towers, and oil rigs against the lightning strikes. These systems stand as an indispensable barrier, enabling operators to take such preventive measures at the proper time to minimize the chance of damages and disruptions capable of essential services.

For example, with the use of automated monitoring and alerting systems operators are immediately notified about potential lightning danger, thus enabling them to quickly react to it and to take measures that are proactive protecting assets and continuity of operations. The adoption and implementation of lightning detection technology into already existing infrastructure and operations will ultimately boost organizations' ability to respond to lightning-related dangers in a more timely and prepared manner.

9 Conclusion

Throughout this whole paper we have mentioned about the current incidents of forest fires related to lightning, the characteristics of lightning, state-of-the-art safety systems, and proposed a system, that can potentially reduce the impact of lightning-caused fire more effectively. Industries and authorities are encouraged to consider the architecture proposed herein and implement it in practical application.

References

1. Caring for lightning-struck trees—Alabama cooperative extension system. https://www.aces.edu/blog/topics/forestry/caring-for-lightning-struck-trees/#:~:text=Lightning%20carries%20up%20to%20100,and%20wood%20fibers%20to%20explode
2. What happens when lightning strikes a tree? Wonderopolis. https://wonderopolis.org/wonder/what-happens-when-lightning-strikes-a-tree?cv=1
3. Moris JV et al (2020) Lightning-caused fires in the Alps: identifying the igniting strokes. Agric For Meteorol 290:107990

4. The understanding of lightning discharge mechanisms—Kingsmill industries. https://kingsmillindustries.com/the-understanding-of-lightning-discharge-mechanism/?cv=1
5. Statistics. National interagency fire center. https://www.nifc.gov/fire-information/statistics
6. Wildfires. Earthdata. https://www.earthdata.nasa.gov/topics/human-dimensions/natural-hazards/wildfires
7. Lightning-caused wildfires. National interagency fire center. https://www.nifc.gov/fire-information/statistics/lightning-caused
8. Bengtsson J, Nilsson SG, Franc A, Menozzi P (2000) Biodiversity, disturbances, ecosystem function and management of european forests. For Ecol Manag 132:39–50
9. Mahomed M, Clulow AD, Strydom S, Savage MJ, Mabhaudhi T (2021) Lightning monitoring and detection techniques: progress and challenges in South Africa. S Afr J Sci 117
10. A comprehensive to NodeMCU ESP8266 pinout. Xecor. https://www.xecor.com/blog/nodemcu-esp8266-pinout
11. Mialdea-Flor I et al (2019) Development of a low-cost IoT system for lightning strike detection and location. Electronics 8:1512
12. AS3935 Franklin lightning sensor™ IC—sciosense. https://www.sciosense.com/as3935-franklin-lightning-sensor-ic/
13. AS3935 datasheet(PDF)—ams AG. https://www.alldatasheet.com/datasheet-pdf/pdf/937220/AMSCO/AS3935.html
14. cmheong's blog: AS3935 lightning detector with I2C and ESP8266 NodeMCU ESP-12E. https://cmheong.blogspot.com/2020/11/as3935-lightning-detector-with-i2c-and.html

Chapter 14
Smart and Effective Healthcare for Diabetic Patients Using ML Techniques

Ashok Kumar Pradhan, Sai Harshitha Dhulipalla, Shaik Tahseen Nishat, and Swetha Ghanta

Abstract Diabetes is a prevalent and enduring condition impacting millions of individuals worldwide this is why early detection is essential for efficient management and intervention. The objective of this article is to build a trustworthy diabetes prediction model using the RF-SVM algorithm and employing ensemble techniques, specifically the stacking technique. The Pima Indians dataset, renowned for its comprehensive clinical and demographic features, is utilized for this study. Our proposed RF-SVM ensemble model trains on a subset of the dataset using stratified cross-validation to increase robustness and generalizability. With an accuracy of 86%, the proposed model successfully predicts diabetes, demonstrating its value in early diagnosis and timely treatment. Feature importance analysis can help us better understand the variables that affect how diabetes develops. This study demonstrates the utility of the RF-SVM ensemble model with the stacking technique for diabetes prediction. The developed approach is effective in identifying patients who are at risk, which improves patient outcomes. Future research initiatives may include merging more datasets and researching advanced machine learning approaches to increase prediction accuracy and increase the model's utility.

Keywords Diabetes prediction · RF-SVM · Ensemble techniques · Stacking · Machine learning · Pima Indians dataset · Feature importance

1 Introduction

Diabetes, a prevalent metabolic condition, poses a significant global health risk due to elevated blood sugar levels. If left untreated, it is a chronic illness that is brought on by elevated blood sugar levels in humans and can impact multiple organs. It aggravates blood vessels, heart disease, kidney problems, damaged nerves, and blindness. Early

A. K. Pradhan (✉) · S. H. Dhulipalla · S. T. Nishat · S. Ghanta
SRM University, Amaravati, Andhra Pradesh, India
e-mail: ashokkumar.p@srmap.edu.in

A. Acharyya et al. (eds.), *Real-World Applications and Implementations of IoT*, Studies in Smart Technologies, https://doi.org/10.1007/978-981-97-8627-5_14

diabetes detection is crucial for reducing complications. The RF-SVM algorithm and ensemble techniques, notably the stacking technique, are used in this study to build an accurate diabetes prediction model. The objective of the work is to better predict the outcomes and identify risk factors by combining the benefits of various models.

Even though clinical and demographic data are routinely employed in traditional diabetes prediction methods, these approaches' poor accuracy compels researchers to explore more advanced alternatives [1]. Machine learning (ML) methods like Random Forest (RF) and Support Vector Machines (SVM) have shown promise in the field of healthcare. These algorithms can precisely forecast the risk of developing diabetes by effectively analyzing complex relationships between a variety of variables. In this study, we apply ensemble techniques with an emphasis on the stacking technique to further improve the prediction performance of the RF-SVM model. By combining the unique advantages of each base model, the stacking technique creates a meta-model that gains from their combined predictive power. This strategy allows us to overcome the limitations and biases inherent in each model, improving the accuracy and dependability of diabetes prediction [2]. This study makes use of the Pima Indians dataset. With the use of this dataset, we train and test the RF-SVM ensemble model to develop a reliable and clinically useful diabetes prediction model.

The objectives of this work are as follows:

- Improve the prediction performance of the RF-SVM model by using ensemble approaches, particularly the stacking technique.
- Analyze the value of features to find important predictors that influence the development of diabetes.

The format of the paper is as follows: Sect. 2 illustrates the contemporary of related works. Sections 3, 4 and 5 discuss the motivation to combine RF and SVM algorithms and the design and implementation of the RF-SVM model. In Sects. 6 and 7, we discuss ensembling techniques and the design and implementation of a Stacking classifier with RF and SVM as base models. Finally, in Sect. 8 we evaluated the proposed model with other algorithms and demonstrated the results followed by Conclusion and References.

2 Literature Review

- Ensemble techniques have grown in popularity in machine learning as they combine predictions from many models to get more precise and dependable results. The authors used six ML algorithms, logistic regression (LR), SVM, k-nearest neighbors (KNN), gradient boost, naive Bayes (NB), and RF on the PIMA Indians dataset to predict preliminary diabetes risk, finding that RF was the best with 80.7% accuracy [3]. They then used a voting algorithm to combine the predictions of three better-performing models: NB, RF, and gradient boosting classifier, achieving an accuracy of 82.0%.

- The ensemble model that the authors have presented combines XGBoost, RF, SVM, and NB. To classify in the direction of heart disease identification, a voting method is employed. The results of the suggested model were 96.75% on the Mendeley Data Center's dataset on cardiovascular disease, 93.39% on IEEE DataPort's complete dataset, and on the Cleveland dataset it is 88.24% [4].
- Saloni et al. introduced an ensemble technique for diabetes mellitus classification and prediction, employing a soft voting classifier. Their method combines three distinct machine learning algorithms—RF, LR, and NB. The ensemble soft voting classifier achieves binary classification with a 71.45% recall, 73.48% precision, 79.04% accuracy, and 80.6% F1 score, as demonstrated on the PIMA diabetes dataset [5].
- In 2020, Alanazi and Mezher conducted an extensive investigation into machine learning methodologies for diabetes prediction. Focusing primarily on two algorithms, RF and SVM, they scrutinized a dataset sourced from a Saudi medical institution. Their analysis concluded that RF exhibited superior accuracy, achieving 87% [6].
- A research study on the Pima Indians Diabetes Dataset and the Practice Fusion Dataset, two diabetes datasets, was carried out by Ramya Akula et al. in 2019. After examining seven distinct machine learning methods, they discovered that, for the Pima Diabetes Dataset, decision trees had the best accuracy at 72%, while neural networks had the best accuracy at 82.5% for the Practice Fusion Dataset. They have also experimented with ensemble and boosting models, yielding even better accuracies. The ensemble model trained on the Practice Fusion Dataset acquired 86% accuracy, while the ensemble model trained on the Pima Indians Diabetes dataset acquired 89.1% accuracy [7].
- An ensemble model for diabetes prediction was presented by Jae-Ho Rhee, Gusti Alfian, Muhammad Syafrudin, and Nabila L. Fitriyani. As the first classification level, they employed decision trees (DT), SVM, and multilayer perceptrons (MLP). They employed linear regression in the second iteration. The accuracy of their suggested model was 92.8% [8].
- Aishwarya Mujumdar and Vaidehi utilized a range of machine learning algorithms, encompassing Bagging, Linear Discriminant Analysis, Ada Boost, Perceptron, LR, KNN, Gradient Boost, Gaussian NB, SVM, RF, DT, Extra Trees, and Ada Boost. They assessed the performance of these models using two datasets: the PIMA Indians dataset and another dataset related to Diabetes. Among these models, LR achieved an accuracy rate of 96% [9].
- Sajida et al. explored the application of J48 decision trees alongside ensemble machine learning methods Bagging and Adaboost for identifying patients as non-diabetic or diabetic considering diabetes risk factors. Their study reveals that the Adaboost outperforms bagging and the J48 DT approach in terms of effectiveness [10].
- The primary goal of the diabetes prediction system created by Orabi et al. is to predict if an individual has diabetes at a given age. By using decision trees, the suggested system is created using the machine learning approach. The system's

ability to accurately anticipate diabetes cases at a specific age through the use of decision trees yielded good results [11].

- Tejas and Pramila decided to construct a prediction model for diabetes with two algorithms: SVM and LR. To achieve better results, the data was pre-processed. With a 79% accuracy, SVM performed better than LR [12].
- Ebenezer et al. employed Deep Learning methods for predicting diabetes. They utilized a Multilayer Feed-Forward Neural Network, with the back-propagation technique. Additionally, they pre-processed the PIMA Indians dataset by normalization to ensure numerical stability, achieving an accuracy rate of 82% [13].
- Deepti et al. utilized DT, SVM, and NB ML models, augmenting performance through tenfold cross-validation on the Pima dataset. The highest accuracy of 76.3% is achieved by NB model [14].
- Yuvaraj et al. proposed a diabetes prediction model on Hadoop-based clusters utilizing RF, DT, and NB algorithms after employing some pre-processing techniques on the dataset. The RF model is the best performing with 94% accuracy [15].
- Naveen et al. employed five distinct machine learning algorithms—SVM, DT, NB, LR, and KNN—to forecast diabetes in the PIMA dataset. The combined machine learning approach achieved a 75% accuracy in predicting diabetics [16].
- Kopitar et al. used three methods to predict diabetes: the RF algorithm, the NB classifier, and KNN. Together with these machine learning algorithms, they also used the LightGBM, Glmnet, and XGBOOST techniques. When it came to diabetes prediction, the XGBOOST fared better than the others. It has an accuracy of 88% [17].

3 Random Forest and SVM Selection

3.1 Random Forest

Due to the following reasons, Random Forest is selected in our proposed work.

- Robustness to Irrelevant Features:
  Random Forest is capable of handling datasets with a large number of features. It assesses each feature's importance based on its contribution to reducing impurity and improving prediction accuracy. By utilizing Random Forest for feature selection, the most relevant features can be retained while discarding less informative or irrelevant ones [18]. This helps in reducing overfitting, enhancing model performance, and simplifying the subsequent SVM classifier's task by focusing on the most discriminative features.
- Capturing Nonlinear Relationships:
  RF, a collection of DTs, is particularly good at capturing intricate and nonlinear interactions between attributes and the dependent variable [19], which is crucial for accurately predicting diabetes. By considering multiple decision trees, Random

Forest provides robust and reliable feature importance rankings, aiding in the selection of relevant features for the subsequent SVM classifier.
- Handling Missing Data and Outliers:
  Random Forest demonstrates robustness to missing data and outliers. It can handle missing values by imputing them based on other available features, preventing data loss during feature selection. Furthermore, the ensemble nature of RF enables it to be less impacted by outliers, as it averages out the impact of independent DTs. This robustness helps maintain the integrity of the feature selection process, even with missing data or outliers in the PIMA India dataset [20].

3.2 SVM

- Handling High-Dimensional Data: The PIMA India dataset contains multiple features, and SVM has been proven to perform well in high-dimensional spaces. It can effectively separate classes in complex data distributions [21].
- Robustness to Outliers: Outliers can have a prominent effect on the ML algorithms' performance. SVM is less affected by outliers due to its reliance on support vectors, which are samples closest to the decision boundary. This property makes SVM more robust and reliable.
- Kernel Trick: Using this, SVM transforms the input space into a feature space of higher dimensions. This allows SVM to handle nonlinear decision boundaries and capture complex relationships between features [22].
- Margin Maximization: The biggest margin between class labels is the hyperplane that SVM seeks to identify. This characteristic promotes generalization and helps avoid overfitting. SVM seeks to achieve better predictive performance on unseen data, by maximizing the margin.

4 RF and SVM Combination

- Complementary Strengths of RF and SVM:
  RF and SVM possess complementary strengths, making their combination highly advantageous. Random Forest excels in identifying relevant features and capturing complex relationships, while SVM is proficient in constructing decision boundaries and making accurate predictions. By combining the two algorithms, we leverage the feature selection power of Random Forest and the discriminative capabilities of SVM, resulting in a robust and accurate prediction model for diabetes [23].
- Improved Generalization and Accuracy:
  The RF-SVM combination leads to improved generalization and accuracy. Random Forest's feature selection process reduces dimensionality and focuses on the most informative features, enhancing the SVM classifier's ability to generalize well

on the reduced feature set. By selecting relevant features and leveraging SVM's capability to find optimal decision boundaries, we can build a more robust diabetes prediction model.

- Addressing Correlated Features:
  The presence of correlated features in the PIMA India dataset can lead to multicollinearity issues and suboptimal performance of the classifiers. By incorporating Principal Component Analysis (PCA) as a feature transformation technique, we aim to mitigate the impact of correlated features, improve interpretability, and enhance the performance of the subsequent SVM classifier.

We present a proposal for improved diabetes prediction using a combination of feature selection with RF, PCA for feature transformation, and classification with SVM algorithms.

5 RF-SVM Model Design and Implementation

5.1 Feature Selection with Random Forest

- Random Forest Training: An RF model is trained on the PIMA India dataset. The RF algorithm builds an ensemble of DTs and evaluates each feature's importance based on their contribution to the overall predictive performance.
- Feature Importance Ranking: The importance scores assigned by RF are used to rank the features in descending order. This ranking helps identify the most relevant features for diabetes prediction.

5.2 PCA for Feature Transformation

- Correlation Analysis: The presence of correlated features, such as Glucose and Outcome, Age and Pregnancies, BMI and SkinThickness, BloodPressure and BMI, and Insulin and Glucose, is identified in the dataset.
- By leveraging PCA to transform the correlated features, we aim to mitigate the potential issues associated with multicollinearity, improve interpretability, and enhance the performance of the subsequent SVM classifier. The resulting principal components will serve as transformed features that can be fed into the SVM classifier for diabetes prediction.

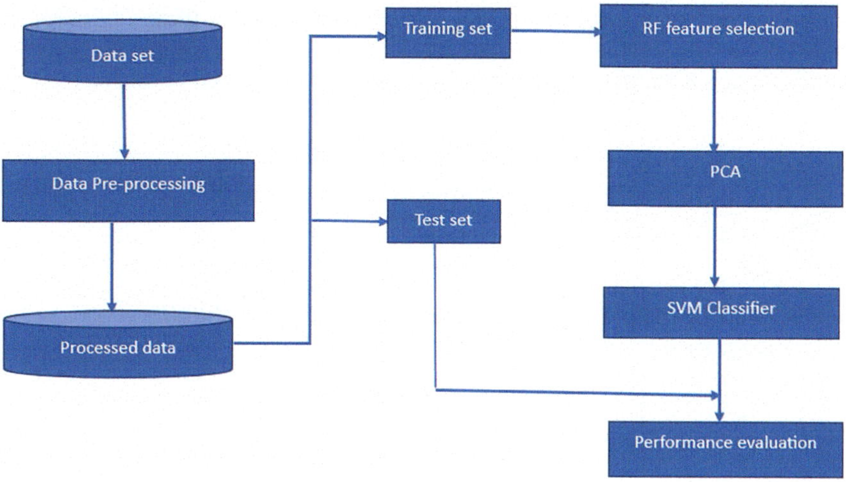

Fig. 1 Workflow diagram of the RF-SVM model

5.3 Classification with Support Vector Machines

- Feature Subset Selection: The top-ranked features from the RF analysis, along with the principal components obtained from PCA, are selected as the feature subset for SVM. This step involves filtering out the less important features and leveraging the transformed features. By eliminating irrelevant features and addressing co-related features, the model becomes more robust to overfitting and noise, leading to improved generalization performance.
- SVM Training: An SVM model is trained using the selected feature subset. SVM seeks an optimal hyperplane to separate the diabetic and non-diabetic instances based on the reduced feature space.
- Prediction Phase (Feature Subset Application:) The trained SVM model uses the selected feature subset including the transformed features obtained from PCA to make predictions for new instances. Only the chosen features are considered during the prediction phase to ensure computational efficiency and improve performance. The workflow of the RF-SVM model can be seen in Fig. 1.

6 Ensemble Techniques for Model Combination: Rationale

In the earlier proposal, the idea was to directly combine Random Forest and SVM for diabetes prediction. However, after careful consideration, it has been decided to adopt Ensembling techniques, as they can improve the prediction accuracy by combining the predictions of multiple models. This can reduce bias and variance [24], improve generalization, and handle noise and outliers more effectively [25]. Ensemble techniques can also adaptively weight the predictions of each model, ensuring that the models with stronger predictive capabilities receive higher weights [26].

7 Stacking Classifier with Random Forest and SVM: Optimal Approach

7.1 Utilizing Stacking Classifier for Improved Diabetes Prediction

Rationale: In the initial proposal, the idea was to directly combine Random Forest and SVM for diabetes prediction by feeding the selected features to SVM. However, after careful consideration, it has been decided to adopt a stacking classifier instead of other ensemble methods like Adaboost and Voting Classifier because

- In AdaBoost, the weak learners are required to be homogeneous, meaning they should be of the same type. This limitation may hinder the utilization of the diverse strengths of Random Forest and SVM. On the other hand, stacking allows for heterogeneous base models, enabling us to combine the distinctive characteristics and diverse perspectives of Random Forest and SVM effectively [26].
- Voting classifiers consider equal weighting for all base models, which may not be optimal if some models are more accurate or reliable than others. The stacking classifier, with its adaptive weighting mechanism, can assign higher weights to models that exhibit better performance on the given dataset, resulting in improved prediction accuracy.

7.2 Base Models: RF and SVM

RF and SVM were chosen as base models for their complementary strengths in handling complex datasets and making accurate predictions. They are also known for their robustness and generalization capabilities, which are important for diabetes prediction models.

7.3 Meta-Learner

To produce more accurate predictions, a meta-learner aggregates the output of several different models. It learns how to best aggregate these predictions by observing their performance on training data. The idea is that by leveraging the strengths of different models, the meta-learner can provide better overall predictions. SVM has been selected as the meta-learner in our ensemble model.

By using SVM as the meta-learner, we can benefit from its capability to capture complex decision boundaries and make accurate predictions based on the outputs of the base models.

7.4 Data Preparation

The PIMA Indians diabetes dataset consists of information collected from Pima Indian women with a minimum age of 21 years old. It consists of 768 rows and 9 columns. Preprocess the PIMA India dataset, including handling missing values, normalizing or standardizing features to a 0 to 1 range, and splitting the data into training and testing sets. We employed a train-test split approach to assess the model. This method involves partitioning the dataset into training and testing sets using a specified test size. The model used a training set for training and a testing set for evaluation. Additionally, we utilized the GridSearchCV method for hyperparameter tuning, dividing the dataset into n folds and assessing the model's classification performance on various data subsets.

7.5 Design and Implementation

- Base Model Training: Train the first base model, Random Forest (RF), and the second base model Support Vector Machines (SVM), on the same training data.
- Base Model Predictions: Generate predictions for new instances by passing them through the base models (RF and SVM), obtaining their individual predictions. Concatenate the individual predictions with the original features and pass them through the trained meta-learner (SVM) to obtain the final prediction for diabetes. Perform hyperparameter tuning on the base models (RF and SVM) and the meta-learner (SVM) using techniques like randomized search or grid search cv to find best-suited parameter settings.
- Meta-Learner Training: Create a new dataset by concatenating the predictions from the RF and SVM base models along with the original features. This dataset will serve as the input to the meta-learner. Train the meta-learner, SVM, using the new dataset. By training the meta-learner on the base model predictions, it can learn to combine and weigh these predictions effectively, taking into account the advantages and disadvantages of each base model. This integration of multiple predictions allows for a more informed and robust final prediction [27].
- Evaluation and Performance Analysis: We evaluate the performance of the stacking classifier by comparing the obtained predictions to the actual labels in the testing data. We used evaluation metrics such as recall, precision, accuracy, and F1 score

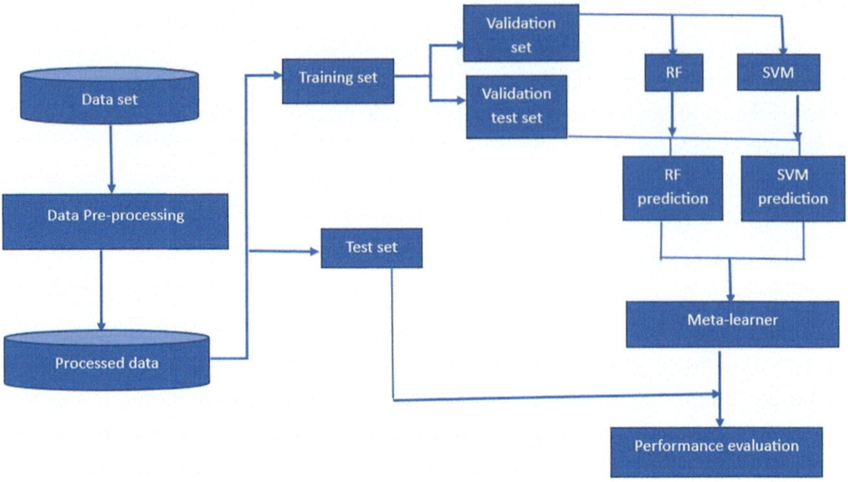

Fig. 2 Workflow diagram of the proposed RF-SVM Meta Learner

to assess the classifier's predictive performance. The workflow of the proposed RF-SVM meta learner can be seen in Fig. 2.

8 Evaluation

To evaluate the predictive performance of the algorithms, we employ the following evaluation metrics:

a. Accuracy:
   The percentage of accurately anticipated cases relative to all instances is known as accuracy. As accuracy gives a broad picture of the algorithm's performance, it is an often utilized metric. That could not be enough, though, if the dataset is unbalanced.
b. Recall (Sensitivity or True Positive Rate):
   The percentage of true positive cases that are accurately anticipated to be positive is called recall. When identifying positive instances is critical, memory is essential. It ensures that fewer true positives are categorized as negatives by reducing false negatives.
c. Precision (Positive Predictive Value):
   The precision metric quantifies the percentage of accurately predicted positive cases among all the cases that are projected to be positive. When there are substantial repercussions from false positives, precision is important. It guarantees that negative cases are not classified as positives by the algorithm.

d. F1 Score:
The F1 score strikes a balance between recall and precision by combining the two into a single metric. When there is an imbalance in the dataset between positive and negative examples, the F1 score is especially helpful. It offers a thorough assessment of the algorithm's performance and takes into account both false positives (FP) and false negatives (FN).

$$\text{Accuracy} = (\text{TP} + \text{TN})/(\text{TP} + \text{TN} + \text{FP} + \text{FN})$$
$$\text{Recall} = \text{TP}/(\text{TP} + \text{FN})$$
$$\text{Precision} = \text{TP}/(\text{TP} + \text{FP})$$
$$\text{F1 Score} = 2 * (\text{Precision} * \text{Recall})/(\text{Precision} + \text{Recall})$$

Using accuracy alone as a performance metric is not efficient, especially with imbalanced datasets. Accuracy does not consider FPs and FNs separately, which may not be desirable in certain scenarios. By employing recall, precision, and F1 score, we gain a more comprehensive understanding of the algorithm's advantages and disadvantages. Recall is critical when the aim is to minimize FNs, such as in predicting diabetes. It ensures that individuals with diabetes are not missed during prediction, enabling timely intervention and treatment. Precision is essential when the consequences of FPs are severe. In diabetes prediction, FPs can lead to unnecessary anxiety and medical interventions. A high precision value indicates that the algorithm is correctly identifying individuals who are likely to develop diabetes. The F1 score provides a balanced evaluation, accounting for both FPs and FNs. It is particularly useful in situations where precision and recall need to be considered simultaneously. Figure 3 illustrates the accuracies of simple ML models, while Fig. 4 highlights those of the ensemble models. Table 1 presents the evaluation metrics of various classifiers.

9 Results

ROC Curves

Fig. 3 Simple ML models

Algorithm	Accuracy
KNN	72.3%
RF	81.2%
SVM	81.2%

Algorithm	Accuracy
RF-SVM	83.1%
AdaBoost Classifier	74%
Stacking Classifier	86.36%

Fig. 4 Ensemble models

Table 1 Comparison of evaluation metrics for different algorithms

Algorithm	Precision	Recall	F1 score
RF	0.704	0.7036	0.7039
KNN	0.566	0.557	0.561
SVM	0.719	0.672	0.694
RF-SVM	0.815	0.553	0.681
AdaBoost classifier	0.6	0.363	0.50
Stacking classifier	0.821	0.675	0.72

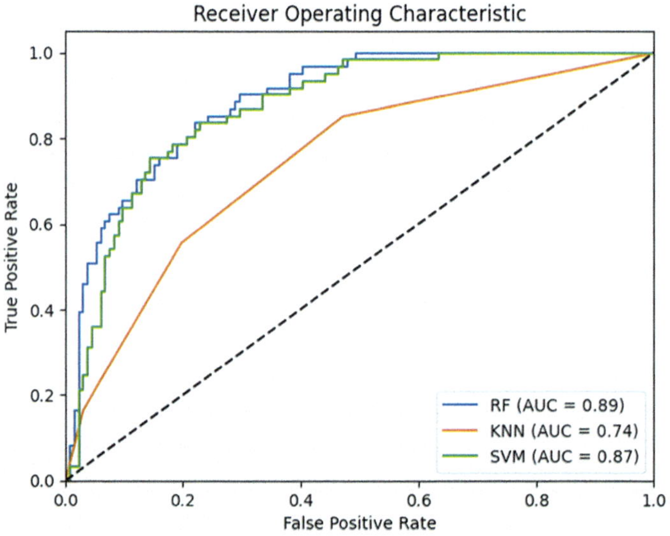

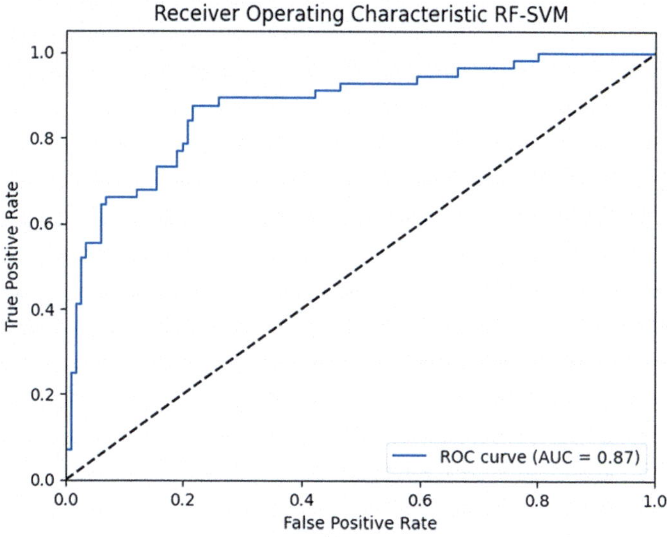
Receiver Operating Characteristic RF-SVM
True Positive Rate
False Positive Rate
ROC curve (AUC = 0.87)
0.0
0.2
0.4
0.6
0.8
1.0

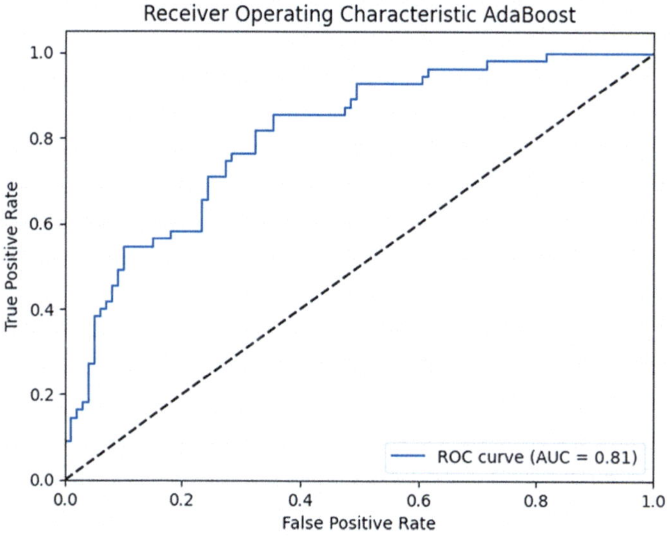
Receiver Operating Characteristic AdaBoost
True Positive Rate
False Positive Rate
ROC curve (AUC = 0.81)
0.0
0.2
0.4
0.6
0.8
1.0

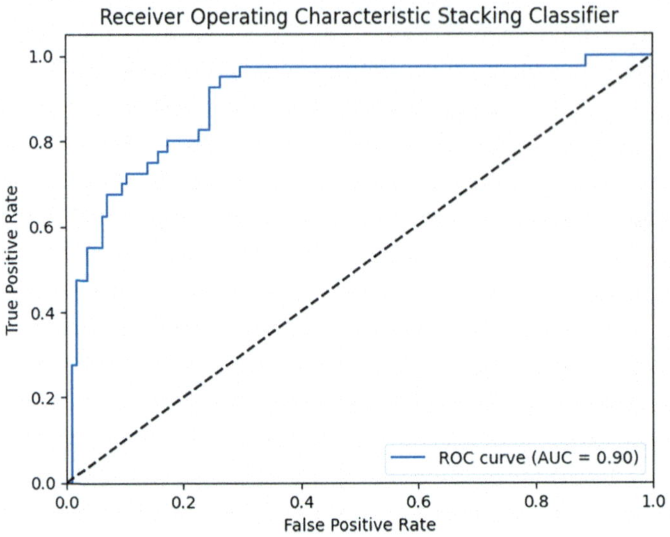

Variation of AUC Scores for All Classifiers

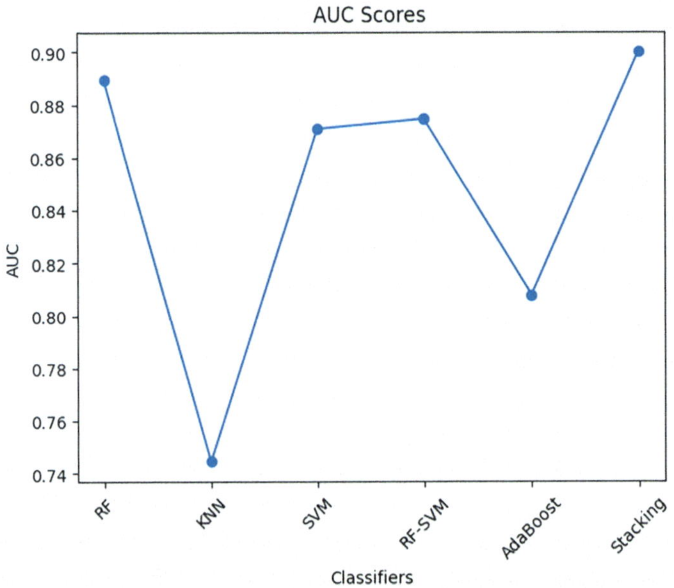

With a 0.821 precision, 0.675 recall, 0.8636 accuracy, and 0.72 F1 score, the stacking classifier performed the best. The second-highest accuracy of 0.831 was attained by the RF-SVM method, while its precision and recall were lower. Diabetes prediction algorithms need to have high precision since it reduce the likelihood of false positive predictions. By doing this, people may be shielded from the harmful effects of incorrect diagnosis. Plotting the true positive rate (TPR) versus the false positive rate (FPR) is known as the ROC curve. The percentage of positive cases that are accurately labeled as positive is known as the TPR, whereas the percentage of negative cases that are mistakenly classified as positive is known as the FPR. They provide a comprehensive analysis of the model's ability to differentiate between negative and positive samples. For the basic models, the ROC curve for RF is closest to the top-left corner, which means that RF is the best algorithm for distinguishing negative and positive samples. The ROC curve for KNN is the furthest away from the top-left corner, which means that it is not the best at distinguishing between positive and negative instances. Similarly, out of all the models, the ROC curve for Stacking is closest to the top-left corner followed by the RF-SVM model, which means that it is the best algorithm for distinguishing between positive and negative instances.

10 Conclusion

The key objective of this study has been to enhance outcome predictions and identify patients at risk by harnessing the strengths of diverse models. In conclusion, this study successfully coupled the RF-SVM algorithm with the stacking ensemble technique to create an accurate predictive model for diabetes. The model's accuracy of 86.36% in predicting diabetes was tested against the Pima Indians dataset. The ensemble technique, which incorporates the benefits of several models, has proven effective in enhancing forecast reliability and accuracy. The results of this study suggest that stacking ensemble models can be effective for predicting diabetes. The use of two base models helped to reduce the risk of bias and variance, and the meta-learner was able to combine the predictions from the base models to make more accurate predictions.

Future research directions may look at additional datasets and incorporate advanced machine learning techniques to further enhance the model's performance and generalizability. However, the RF-SVM ensemble model with a stacking approach is an effective tool for predicting diabetes and promotes better patient outcomes and preventive healthcare practices.

Acknowledgements The authors are grateful to the University Research Grant-Central Facility, SRM AP, for funding this work, Sanction number SRMAP/URG/CF/2023-24/045.

References

1. Kibria HB, Nahiduzzaman M, Goni MOF, Ahsan M, Haider J (2022) An ensemble approach for the prediction of diabetes mellitus using a soft voting classifier with an explainable AI. Sensors 22(19). https://www.mdpi.com/1424-8220/22/19/7268
2. Mahajan P, Uddin S, Hajati F, Moni MA (2023) Ensemble learning for disease prediction: a review. Healthcare 11:1808

3. Mushtaq Z, Ramzan M, Ali S, Baseer S, Samad A, Husnain M (2022) Voting classification-based diabetes mellitus prediction using hypertuned machine-learning techniques. Mobile Inf Syst 2022:1–16
4. Doppala BP, Bhattacharyya D, Janarthanan M, Baik N (2022) A reliable machine intelligence model for accurate identification of cardiovascular diseases using ensemble techniques. J Healthc Eng 2022:03
5. Kumari S, Kumar D, Mittal M (2021) An ensemble approach for classification and prediction of diabetes mellitus using soft voting classifier
6. Alanazi A, Mezher M (2020) Using machine learning algorithms for prediction of diabetes mellitus 1–3
7. Akula R, Nguyen N, Garibay I (2019) Supervised machine learning based ensemble model for accurate prediction of type 2 diabetes. In: SoutheastCon, pp 1–8
8. Fitriyani NL, Syafrudin M, Alfian G, Rhee J (2019) Development of disease prediction model based on ensemble learning approach for diabetes and hypertension. IEEE Access 7:144,777–144,789
9. Mujumdar A, Vaidehi V (2019) Diabetes prediction using machine learning algorithms. Procedia Comput Sci 165:292–299
10. Perveen S, Shahbaz M, Guergachi A, Keshavjee K (2016) Performance analysis of data mining classification techniques to predict diabetes. Procedia Comput Sci 82:115–121
11. Orabi KM, Kamal YM, Rabah TM (2016) Early predictive system for diabetes mellitus disease. In: Advances in data mining. Applications and theoretical aspects: 16th industrial conference, ICDM, New York, NY, USA, July 13–17, 2016. Proceedings 16. Springer, pp 420–427
12. Joshi TN, Chawan PM (2018) Logistic regression and SVM based diabetes prediction system. Int J Technol Res Eng 5:4347–4350
13. Olaniyi EO, Adnan K (2014) Onset diabetes diagnosis using artificial neural network. Int J Sci Eng Res 5(10):754–759
14. Sisodia D, Sisodia DS (2018) Prediction of diabetes using classification algorithms. Procedia Comput Sci 132:1578–1585
15. Yuvaraj N, SriPreethaa K (2019) Diabetes prediction in healthcare systems using machine learning algorithms on Hadoop cluster. Clust Comput 22(Suppl 1):1–9
16. Kishore GN, Rajesh V, Reddy AVA, Sumedh K, Reddy TRS (2020) Prediction of diabetes using machine learning classification algorithms. Int J Sci Technol Res 9(1):1805–1808
17. Kopitar L, Kocbek P, Cilar L, Sheikh A, Stiglic G (2020) Early detection of type 2 diabetes mellitus using machine learning-based prediction models. Sci Rep 10(1):11981
18. Chen R-C, Dewi C, Huang S, Caraka R (2020) Selecting critical features for data classification based on machine learning methods. J Big Data 7:26
19. Auret L, Aldrich C (2012) Interpretation of nonlinear relationships between process variables by use of random forests. Miner Eng 35:27–42
20. Tang F, Ishwaran H (2017) Random forest missing data algorithms
21. Mathur A, Foody G (2008) Multiclass and binary SVM classification: implications for training and classification users. Geosci Remote Sens Lett IEEE 5:241–245
22. Sheykhmousa RM, Mahdianpari M (2020) Support vector machine vs. random forest for remote sensing image classification: a meta-analysis and systematic review. IEEE J Select Top Appl Earth Obs Remote Sens
23. Egala BS, Pradhan AK, Dey P, Badarla V, Mohanty SP (2023) Fortified-chain 2.0: intelligent blockchain for decentralized smart healthcare system. IEEE Internet Things J 10(14):12,308–12,321
24. Cawley GC, Talbot NLC (2010) On over-fitting in model selection and subsequent selection bias in performance evaluation. J Mach Learn Res 11(70):2079–2107. http://jmlr.org/papers/v11/cawley10a.html
25. Galar M, Fernandez A, Barrenechea E, Bustince H, Herrera F (2012) A review on ensembles for the class imbalance problem: bagging-, boosting-, and hybrid-based approaches. IEEE Trans Syst, Man, Cybern Part C (Appl Rev) 42(4):463–484

26. Doganer A (2021) Different approaches to reducing bias in classification of medical data by ensemble learning methods. Int J Big Data Anal Healthc 6:15–30
27. Ghasemieh A, Lloyed A, Bahrami P, Vajar P, Kashef R (2023) A novel machine learning model with stacking ensemble learner for predicting emergency readmission of heart-disease patients. Decis Anal J 7:100242. https://www.sciencedirect.com/science/article/pii/S2772662223000826

Chapter 15
Comparative Study of Feature Selection Algorithms for Heart Disease Prediction

Preetam Vallabhaneni, Aditya Dudugu, Harishitha Chowdary Alapati, Sravya Meda, and Ashok Kumar Pradhan

Abstract The heart plays a vital role in living organisms. Heart-related disorders are more difficult to diagnose and predict, and errors in diagnosis and prognosis can have fatal consequences. A subfield of artificial intelligence, machine learning uses training from natural occurrences to predict any type of event with minimal support. In this work, we estimate the predictive power of machine learning algorithms that use several feature selection strategies to forecast heart disease. These algorithms include Select k best, Mutual Information Score, Recursive Feature Elimination (RFE), Support Vector Machine (SVM) estimator, and cross-validation. Random forest and gradient boosting were the predictive models used on the Framingham dataset for training and testing. This study uses the SMOTE technique for class imbalance problems. In our research, we got the highest accuracy for select k best feature selection modeled using random forest with an accuracy of 95.61% and gradient boosting with an accuracy of 95.43% considering hyperparameters.

Keywords SMOTE · RFE · Select k best · Mutual information score · Heart disease · SVM · Random Forest · Gradient boosting

1 Introduction

Most of the diseases nowadays are related to the heart, so it is necessary to predict heart diseases. To this end, a comparative study in this field is required. Today, a majority of patients are dying because their diseases were discovered too late due to a lack of instrument accuracy, so it is vital to learn about more effective algorithms for disease prediction.

Labeled training data are necessary for supervised learning techniques, and in classification issues, every data sample is part of a predefined class or category. Class imbalance arises when one class, the minority group, has substantially less samples

P. Vallabhaneni · A. Dudugu · H. C. Alapati · S. Meda · A. K. Pradhan (✉)
SRM University, Amaravati, Andhra Pradesh, India
e-mail: ashokkumar.p@srmap.edu.in

A. Acharyya et al. (eds.), *Real-World Applications and Implementations of IoT*, Studies in Smart Technologies, https://doi.org/10.1007/978-981-97-8627-5_15

than the other class, the majority group, in a binary classification issue involving data samples from two groups. The problem of detecting disease in medical diagnostics, when most patients are healthy and disease detection is of greater importance, is a well-known example of class unbalanced machine learning. It may be quite challenging to learn from these unbalanced datasets, particularly when working with large amounts of data. To resolve this problem, there are many efficient techniques a few of them being resampling, algorithmic, and synthetic data generation techniques.

Resampling approaches modify the distribution of classes within the training set. Increasing the minority class occurrences by over-sampling usually entails creating fake or duplicate data. By doing this, class proportions are balanced and the model is kept from favoring the majority class. By lowering the quantity of majority class samples, undersampling lessens the class's dominance throughout training. Algorithmic approaches are to properly control class imbalance by altering the learning algorithms. For example, cost-sensitive learning prioritizes minority class occurrences during training by allocating distinct misclassification costs to classes according to their imbalance level. Synthetic data creation approach works by interpolating between known cases, and generates synthetic samples for the minority class [1]. SMOTE helps generalize the model to unknown data and reduces the danger of overfitting that comes with using basic oversampling strategies. SMOTE creates synthetic samples by taking into account the immediate neighborhood of minority class occurrences. SMOTE, as opposed to simple oversampling, lowers the possibility of overfitting and enhances the model's ability to generalize. It is a popular option in many unbalanced categorization settings because of its efficacy in addressing the issue of class imbalance while preserving the variety and integrity of the minority class.

It is necessary to handle and interpret high-dimensional and heterogeneous data for diagnosing different diseases with a computer-based system [2]. High-dimensional data typically causes overfitting of the model and lengthens the training period. Feature selection is a dimension reduction strategy that involves removing characteristics that are redundant or unnecessary and have no discernible effect on classifier performance in order to reduce a set of quite large data features to a manageable set. Recently, a number of efficient feature selection techniques have been created to lessen the effects of dimensionality. Feature selection algorithms fall into one of three categories: semisupervised, supervised, or unsupervised. Using labeled data, the first and most common method for picking features is supervised feature selection. Three approaches—filter, wrapper, and embedding approaches—are used in supervised feature selection techniques [3].

The Filter Method selects characteristics based on statistical metrics. This technique selects the features as a preprocessing step and is independent of the learning algorithm. Using several metrics through ranking, the filter approach removes duplicated columns and extraneous features from the model. Using filter techniques has the advantages of having a minimal computing time and not overfitting the data. Wrapper approaches entail training and testing a model iteratively in order to systematically evaluate subsets of characteristics. This procedure comprises choosing a subset of characteristics, using that subset to train a model, and then assessing the model's performance. Typically, a predetermined statistic like accuracy, AUC, or F1-score is used to evaluate performance. After making modifications to the chosen subset, the procedure is repeated until a stopping criterion—such as a maximum number of iterations or a specified performance level—is satisfied. This iterative process optimizes the model's prediction power and improves generalization by assisting in the identification of the most informative characteristics. Feature selection is included into the model training process directly using embedded techniques. These techniques do away with the requirement for independent feature selection procedures by choosing the most pertinent characteristics during model training. Through this integration, the model is able to determine which characteristics are most useful for making predictions, which results in a more effective and efficient feature selection process. When it comes to models with inbuilt feature selection processes, including decision tree-based algorithms like Random Forests and Gradient Boosting Machines, or LASSO regularization in linear models, embedded techniques are very helpful. Embedded techniques enhance prediction performance and simplify the modeling process by taking feature significance into account during training.

In this research, we aim to investigate the effectiveness of machine learning algorithms in predicting heart disease. To achieve this goal, we employed a variety of techniques, including random forest and gradient boosting, because of their stability and ability to handle complex data [4]. Since random forests reduce overfitting problems in individual decision trees, they are good at handling noisy medical data and perform well in tests predicting cardiac disease. Furthermore, because they may choose subsets of attributes for each tree, random forests are useful in managing high-dimensional datasets that are often linked to the prediction of heart disease while retaining predictive accuracy. They are especially well-suited to modeling the complex interactions found in patient data because of their capacity to represent nonlinear correlations between different risk factors and heart disease. On the other hand, because gradient boosting like XGBoost and LightGBM can create robust predictive models by successively lowering mistakes, they are also well-suited for the prediction of heart disease. By concentrating on the most difficult cases, they incrementally improve forecasts, which can be vital in spotting tiny patterns suggestive of cardiac disease. Gradient boosting techniques can efficiently use a variety of patient data for precise predictions since they are skilled at managing both numerical and categorical variables that are frequently seen in medical datasets. Gradient boosting models also offer comprehensible estimates of feature relevance, which help medical professionals pinpoint the major risk factors that lead to heart disease. By utilizing ensemble learning techniques, random forest and gradient boosting algorithms provide strong and efficient methods for predicting cardiac disease.

2 Literature Review

Several existing systems and technologies have been developed for heart disease prediction, showcasing advancements in medical and clinical aspects. These systems employ various methodologies, including machine learning and deep learning techniques, to interpret coronary heart disease. Current systems usually start by gathering a wide variety of patient data from several sources, such as genetic information, medical imaging, laboratory testing, electronic health records (EHRs), and lifestyle factors. In order to manage missing values, standardize numerical characteristics, encode categorical variables, and eliminate noise or outliers, this data has undergone preprocessing. By reducing dimensionality and extracting pertinent features, feature engineering strategies can improve the models' prediction ability. For the purpose of predicting cardiac disease, several machine learning algorithms—such as logistic regression, decision trees, random forests, support vector machines (SVM), gradient boosting machines (GBM), and neural networks—are assessed.

Agarwal and Pal [5] offered a strategy for selecting features sequentially in order to identify mortality events in heart disease patients while they are receiving therapy and determine which features are most important. Many machine learning techniques are used, such as DT, GBC, SVM, LDA, KNN, and RF. To further verify the outcomes of the SFS algorithm, the confusion matrix, receiver operating characteristic curve, precision, recall rate, and F1-score are also produced. According to the experimental results, the random forest classifier obtains an accuracy of 86.67% when using the sequential feature selection technique.

Gao et al. [6] proposed a model for predicting cardiac disease that combines feature extraction techniques (PCA and LDA) with ensemble approaches (boosting and bagging). Using a subset of characteristics from the Cleveland heart disease dataset, the authors tested five classifiers (SVM, KNN, RF, NB, and DT) with ensemble approaches (bagging and boosting). The trials' findings showed that the bagging ensemble learning approach with feature extraction from DT and PCA produced the best results.

Senan et al. [7] created a diagnostic model with 24 features based on a dataset of 400 patients to detect chronic renal disease. To determine which traits were most important, recursive feature elimination, or RFE, was employed. This paper uses the random forest, decision tree, support vector machine (SVM), and k-nearest neighbors (KNN) classification techniques. Every categorization technique performed exceptionally well.

Alizadehsani et al. [8] applied machine learning for coronary artery disease, and broke down the primary methodologies into dataset analysis, weight research, implementation techniques, and machine learning (ML). In this study, machine learning classifiers were used. The random forest classifier outperformed every classification model that was evaluated before the hepatitis investigation.

KarenGárate-Escamila et al. [9] proposed a hybrid dimensionality reduction technique combining Chi-square and principal component analysis (CHI-PCA) to predict heart disease. Their study was conducted on three different datasets: Hungarian,

Cleveland, and Hungarian-Cleveland datasets gathered from the UCI Machine Learning Repository. The performance of the suggested technique was evaluated with five different classifiers: random forests, gradient-boosted tree, decision tree, multilayer perceptron, and logistic regression. Chi-square and principal component analysis (CHI-PCA) using random forests (RF) showed the most remarkable accuracy, at 98.7% for the Cleveland dataset, 99.0% for the Hungarian dataset, and 99.4% for the Cleveland-Hungarian (CH) dataset, respectively.

Spencer et al. [10] conducted experiments on four frequently used heart disease datasets using four different feature selection techniques: principal component analysis, Chi-squared testing, ReliefF, and symmetrical uncertainty. As noted by the authors, the benefits of feature selection differ depending on the machine learning approach employed for the cardiac datasets. For example, one of the most accurate models discovered had an accuracy of 85.0%, a precision of 84.73%, and a recall of 85.56% when Chi-squared feature selection was combined with the BayesNet classifier.

3 Methodology

In this paper, a comparative study of feature selection algorithms is proposed, including two filter-based methods (Select K Best and Mutual Information Score) and two wrapper-based methods (Recursive Feature Elimination and Recursive Feature Elimination with Cross-Validation) [1]. The Framingham dataset was collected and preprocessing operations were performed. To solve the class imbalance problem SMOTE technique is applied. Selected the best features using different algorithm. Split the dataset into training data and test data. The models were trained, and labels were predicted for the test data. The workflow of the proposed methodology is illustrated in Fig. 1.

3.1 Dataset Collection and Preprocessing

The Framingham dataset was collected from the Kaggle website. The dataset contains 4,241 records, 14 features, and a target variable tenyearCHD. The attributes are as follows:

- Sex: male(0) or female(1)
- Age: Patient's age
- PresentSmoker: Whether patient is present smoker or not
- CigsPerDay: the average number of cigarettes smoked by the individual in a given day
- BPMeds: the presence or absence of blood pressure medicines in the patient
- PrevalentStroke: the patient's history of stroke, whether or not it occurred before

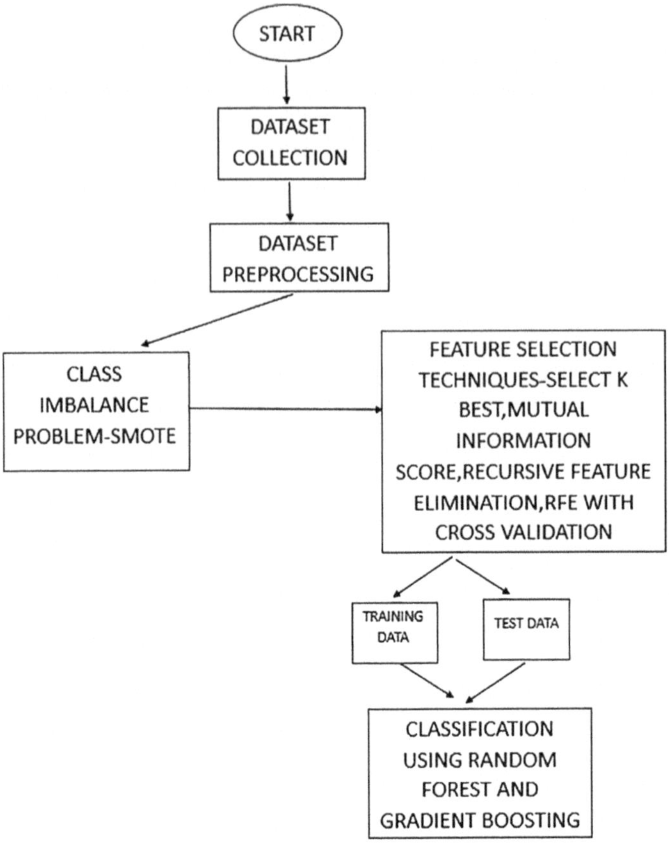

Fig. 1 Workflow

- PrevalentHyp: the degree of hypertension in the patient
- Diabetes: the presence or absence of diabetes in the patient TotChol: total cholesterol level in the patient
- SysBp: Systolic Blood Pressure
- DiaBp: Diastolic Blood Pressure
- BMI: Body mass Index
- Heartrate: the rate of heartbeat
- Blood sugar: glucose level ten-year risk of heart attacks
- CHD: “0” denotes “No” while “1” denotes “Yes.”

Categorical features: male, education, currentSmoker, BPMeds, prevalentStroke, prevalentHyp, and diabetes.

Numerical features: cigsPerDay, totChol, sysBP, diaBP, BMI, heartRate, and glucose.

The missing values for numerical features are handled by replacing them with mode values. This method ensures that the dataset remains complete. The rows containing missing values for numerical features are removed; this approach maintains data integrity and avoids introducing potentially biased imputations. Outliers involving identifying and excluding data points that deviate significantly from the rest of the dataset are removed.

3.2 SMOTE Technique—For Class Imbalance Problem

From the feature space, data points of minority class are selected, SMOTE generates new sample at a location along the line that connects the instances in the features space, instead of resampling which just duplicates the data SMOTE creates new data points avoiding overfitting.
Working Procedure of SMOTE:
Choose the Minority Class: Choose the class that represents the minority in dataset.
Identification of Nearest Neighbors: Determine the k-nearest neighbors of each minority class instance. The parameter that must be provided in advance is the number of neighbors to take into account, represented by k.
Creation of Synthetic Samples: Choose one of the k-nearest neighbors at random for each instance of the minority class. Next, create a fictitious instance along the stretch of line that links the chosen neighbor and the minority instance. In the feature space, this generates a brand-new synthetic instance.
Model Training: After creating synthetic examples, train the machine learning model using the balanced dataset, which consists of both synthetic and original minority class cases.

From Fig. 2, it is evident that the classes are imbalanced. After applying SMOTE, the classes are balanced, as shown in Fig. 3.

3.3 Feature Selection Algorithms

Identifying the most important features from the dataset, this plays a critical role in enhancing the effectiveness and performance of machine learning models. These techniques aid in dimensionality reduction, overfitting mitigation, and improved interpretability of the model by choosing a subset of features that provide the most contribution to the prediction job while eliminating unnecessary or redundant features. Techniques used in this research are as follows.

Fig. 2 Before SMOTE

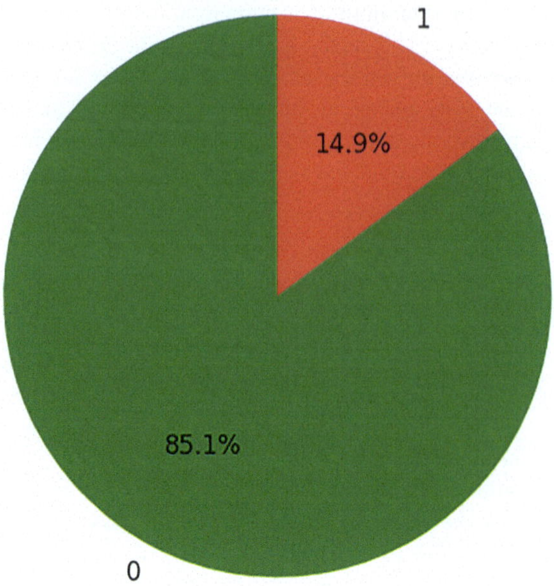

Fig. 3 After SMOTE

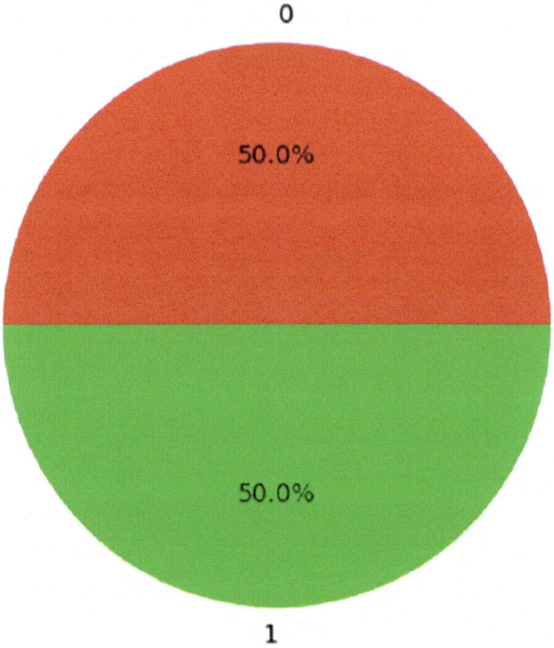

3.3.1 Filter Methods

The SelectKBest Method: Each feature is given a rank based on its univariate statistical significance in relation to the target variable using a straightforward feature selection method called SelectKBest. After analyzing each characteristic independently, it determines the top K traits with the highest scores. This technique significantly reduces dimensionality while retaining the most relevant properties for prediction, which makes it ideal for large feature datasets.

Mutual Information Score: Mutual Information Score is a useful statistic for feature selection tasks since it can be used to measure the degree of dependency between two random variables. It calculates how much knowledge can be gleaned about one variable by using the other. Mutual Information Score evaluates each feature's information gain in relation to the target variable in order to determine its relevance in the context of heart disease prediction. Higher mutual information score features are regarded as more predictively informative.

3.3.2 Wrapper Methods

RFE: The iterative feature selection RFE works by deleting the least important features from the dataset one at a time until the ideal subset of features is found. It evaluates each feature's relevance by using the coefficients or feature importance scores from a selected machine learning model. RFE improves model performance by effectively finding intricate correlations between features and the target variable.

Recursive Cross-Validation Feature Elimination (RFECV): The aim of RFECV is to use accurate number of features while taking model variation into consideration. It does this by combining the concepts of RFE with cross-validation. Using cross-validation, this algorithm selects subsets of features iteratively and assesses their performance. Taking into account the consistency of feature rankings across multiple folds, RFECV ensures robust feature selection and generalization to unseen data.

3.4 Predictive Models

3.4.1 Random Forest

Random Forest is a classifier that generates many decision trees on different dataset subsets and averages them to increase the dataset's predicted accuracy.
Random Data Sampling: A bootstrapping technique will be used for selecting a random subset from the training data. Different subsets are used to train different trees in the forest.

Random Selection of features: Random set of features are taken for splitting each decision at a node. By introducing unpredictability, this contributes to the decorrelation of the forest's trees, increasing their diversity.
Creating Decision Trees: Nodes are divided recursively according to the best feature and split criterion in order to create decision trees. Till a condition is satisfied such as maximum tree depth or minimum amount of leaf nodes, this process continues.
Aggregation of Predictions: The most frequent class label, or mode, among all the trees' predictions is used as the final prediction in classification tasks. The average prediction made by each tree is calculated for regression tasks.

3.4.2 Gradient Boosting

Gradient boosting is a potent boosting procedure that turns several weak learners into strong learners. The approach calculates the gradient of the loss function in relation to the current ensemble's predictions for each iteration, and then trains a new weak model to minimize this gradient. Next, the new model's predictions are included in the ensemble, and the procedure is continued until a stopping requirement is satisfied.
Weak Learners: Gradient boosting starts by fitting a weak learner to the data, typically a decision tree with only a few nodes (a shallow tree). This tree makes predictions, but its accuracy might not be high.
Residuals: The residuals, or the disparities between the predicted and actual values in the training data, are then computed. The mistakes made by the model are represented by these residuals.
Next Tree Fit to Residuals: Next, the residuals are fitted to the next weak learner. This learner concentrates on and attempts to fix the mistakes committed by the first tree.
Iterative Process: This procedure is repeated, concentrating on the residuals of the total predictions made by all of the preceding trees in each new tree. Every tree attempts to fix the mistakes made by the group of trees it came before.
Combining Trees: The ultimate forecast is created by combining the predictions from each tree.

3.5 Model Training

The Random Forest and Gradient Boosting models predict underlying patterns from the train dataset. Encoding is performed to train and test the dataset which trains the model to extract patterns and validates from the test dataset which enables it to make accurate predictions.

4 Results and Discussions

The results include an evaluation of feature selection techniques Select k best, Mutual Information Score, Recursive Feature Elimination, and RFE using cross-validation modeled using Random Forest and Gradient Boosting. Notably, analysis of the modeling outcomes consistently demonstrated that with respective to feature selection algorithm Random Forest performed better compared to Gradient Boosting. Select k best technique with Random Forest modeling gave an accuracy of 95.61%. This robust performance underscores the significant utility of Random Forest and Gradient Boosting in heart disease prediction. Furthermore, comparative analysis revealed that the filter-based feature selection algorithms achieved maximal accuracy compared to wrapper methods. Since the dataset is preprocessed and made majority, minority class balance achieved a good accuracy.

4.1 Equations

Accuracy: The proportion of correctly classified instances among all instances.
F1-Score: The harmonic mean of precision and recall.
Precision: The proportion of true positive predictions among all positive predictions.
Recall: The proportion of true positive predictions among all actual positive instances.
The formulas for the evaluation metrics are provided in Fig. 4.

The following key performance metrics are used to evaluate classification models: F1-score, accuracy, precision, and recall. Accuracy, a measure of the proportion of correctly detected occurrences in the dataset, provides a broad assessment of the model's prediction ability. Precision quantifies the percentage of true positive predictions among all positive predictions to determine how well the model can identify relevant events. Recall measures how many true positive predictions there are among all actual positive instances, indicating how well the model captures all

Fig. 4 Formulas

$$Precision = \frac{TP}{TP + FP} \qquad Recall = \frac{TP}{TP + FN}$$

$$Accuracy = \frac{TP + TN}{TP + FP + FN + TN}$$

$$F1\ Score = 2 \times \frac{Precision \times Recall}{Precision + Recall}$$

relevant events. On the other hand, the F1-score considers both recall and accuracy to provide an equitable evaluation of a model's performance, particularly in cases when class distributions are not balanced.

4.2 Why Do Filter Methods Perform Better Than Wrapper Methods?

Filter methods are computationally less expensive compared to wrapper methods. The Framingham dataset is one of the large datasets for heart disease prediction making filter methods to take advantage over feature relevance, learning fast, and be more scalable. Also, wrapper method eliminated the target variable tenyearCHD while feature selection led to the main drawback. In the correlation matrix generated, education mostly had negative values which indicates that it is less relevant for the prediction but both the wrapper methods picked that feature in the subset as an important feature. The learning algorithm of choice has no bearing on filter techniques. Before using machine learning model, they offer insights into the dataset by evaluating features according to their unique qualities.

5 Conclusion

The research offers a thorough understanding for the importance of feature selection algorithms in bioinformatics and biomedical research. Two well-known machine learning models Random Forest and Gradient Boosting performances were observed. The research finds ideal methodologies that maximizes predicted accuracy for heart disease prediction. Compared to the wrapper methods, filter methods performed better. Random forest classifier paired with Select k best performed with an accuracy of 95.61% and Gradient Boosting paired with mutual information score performed with an accuracy of 95.4% under hyperparameter tuning. However, despite the impressive accuracy achieved using hyperparameter tuning, individual accuracy was 90.39% as shown in Fig. 5. This analysis underscores the importance of considering not only classifiers and feature selection techniques but also the specific hyperparameters considered.

Feature Selection Method	Classifier	Hyperparameter Tuning	Accuracy	Precision	Recall	F1-Score
Select K Best	Random Forest	Yes	95.61	97	94	96
Select K Best	Gradient Boosting	Yes	95.43	97	93	95
Select K Best	Random Forest	No	90.39	96	84	90
Select K Best	Gradient Boosting	No	71.46	73	68	70
Mutual Information Gain	Random Forest	No	89.98	95	84	89
Mutual Information Gain	Gradient Boosting	No	71.42	73	68	70
Recursive Feature Elimination	Random Forest	No	44.10	43	39	40
Recursive Feature Elimination	Gradient Boosting	No	49.44	45	40	41
Rfe with Cross Validation	Random Forest	No	44.99	43	39	40
Rfe with Cross Validation	Gradient Boosting	No	20.65	17	12	14

Fig. 5 Performance metrics including Accuracy, Precision, Recall, and F1-Score

Acknowledgements The authors are grateful to the University Research Grant-Central Facility, SRM AP, for funding this work, Sanction number SRMAP/URG/CF/2023-24/045.

References

1. Jadwal PK, Jain S, Pathak S et al (2022) Improved resampling algorithm through a modified oversampling approach based on spectral clustering and SMOTE. Microsyst Technol 28:2669–2677
2. Balachandran PV, Xue D, Theiler J, Hogden J, Gubernatis JE, Lookman T (2018) Importance of feature selection in machine learning and adaptive design for materials. In: Lookman T, Eidenbenz S, Alexander F, Barnes C (eds) Materials discovery and design. Springer Series in Materials science, vol 280. Springer, Cham

3. Khattak IU, Khan A, Khan FH, Gani A, Shiraz M (2022) A novel feature selection method for classification of medical data using filters, wrappers, and embedded approaches. Hindawi Complex 2022
4. Ahmad GN, Ullah S, Algethami A, Fatima H, Akhter SMH (2022) Comparative study of optimum medical diagnosis of human heart disease using machine learning technique with and without sequential feature selection. IEEE Access 10:23808–23828
5. Aggrawal R, Pal S (2020) Sequential feature selection and machine learning algorithm-based patient's death events prediction and diagnosis in heart disease. SN Comput Sci 1(6)
6. Gao X-Y, Ali AA, Hassan HS, Anwar EM (2021) Improving the accuracy for analyzing heart diseases prediction based on the ensemble method. Complexity 2021. Article ID 6663455, 10 pages
7. Senan EM, Al-Adhaileh MH, Alsaade FW et al (2021) Diagnosis of chronic kidney disease using effective classification algorithms and recursive feature elimination techniques. J Healthc Eng 2021, Article ID 1004767, 10 pages
8. Alizadehsani R, Abdar M, Roshanzamir M et al (2019) Machine learning-based coronary artery disease diagnosis: a comprehensive review. Comput Biol Med 111, Article ID 103346
9. KarenGárate-Escamila A, Hassani AE, Andrès E (2020) Classification models for heart disease prediction using feature selection and PCA. Inf Med Unlocked 19, Article ID 100330
10. Spencer R, Thabtah F, Abdelhamid N, Thompson M (2020) Exploring feature selection and classification methods for predicting heart disease. Digit Health 6, Article ID 2055207620914777

The manufacturer's authorised representative in the EU is Springer Nature Customer Service Centre GmbH, Europaplatz 3, 69115 Heidelberg, Germany. If you have any concerns regarding our products, please contact ProductSafety@springernature.com

Printed and bound by CPI Group (UK) Ltd, Croydon, CR0 4YY
01/10/2026
02225926-0005